PHILOPONUS
On Aristotle's Physics 5-8

with

SIMPLICIUS
On Aristotle on the Void

PHILOPONUS

On Aristotle's *Physics 5-8*

with

SIMPLICIUS

On Aristotle on the Void

Translated by
Paul Lettinck and
J. O. Urmson

Cornell University Press
Ithaca, New York

For information address Cornell University Press,
Sage House, 512 East State Street, Ithaca, New York 14851-0250.

First published 1994 Cornell University Press.

Library of Congress Cataloging-in-Publication Data

Philoponus, John, 6th cent.
On Aristotle's Physics 5-8 / Philoponus. With On Aristotle on the void / Simplicius ; translated by Paul Lettinck and J.O. Urmson.
p. cm. – (Ancient commentators on Aristotle)
Includes bibliographical references and index.
ISBN 0-8014-3005-4
1. Science, Ancient. 2. Physics–Early works to 1800.
I. Lettinck, Paul. II. Urmson, J. O. III. Simplicius, of Cilicia.
On Aristotle on the void. 1994. IV. Title. V. Series.
Q151.P48 1994
500–dc20 93-31609

Acknowledgments

The present translations have been made possible by generous and imaginative funding from the following sources: the National Endowment for the Humanities, Division of Research Programs, an independent federal agency of the USA; the Leverhulme Trust; the British Academy; the Jowett Copyright Trustees; the Royal Society (UK); Centro Internazionale A. Beltrame di Storia dello Spazio e del Tempo (Padua); Mario Mignucci; Liverpool University. The editor wishes to thank Dr Fritz Zimmermann and Dr Peter Lautner for their comments on the translations, Elaine Miller for additional editing of the Philoponus translation, and Ian Crystal, Paul Opperman and Dirk Baltzly for their help in preparing the volume for press.

Printed in Great Britain

Contents

Preface *Richard Sorabji* vii

Philoponus: On Aristotle *Physics* 5-8
translated by Paul Lettinck

Translator's Note 2
Introduction 3
Translation 19
Notes 137
English-Arabic-Greek Glossary 151
Index of Names and Subjects 154

Simplicius: On Aristotle on the Void
translated by J.O. Urmson
introduction and notes by Peter Lautner

Introduction 159
Translator's Note 165
Textual Emendations 166
Translation 167
Notes 223
English-Greek Glossary 237
Greek-English Index 242
Subject Index 263

Preface

Richard Sorabji

This volume presents works on Aristotle's *Physics* by the two rivals, Philoponus and Simplicius, both Neoplatonists of the sixth century AD, the first a Christian, the second a pagan. Philoponus' text, lost in Greek except for fragments, is here translated for the first time from the Arabic. The Arabic, included in an Arabic translation of Aristotle, is a paraphrase of Philoponus' commentary on *Physics* Books 5 to 7, with two comments on the final book, 8.

The Simplicius text comes from his huge commentary on *Physics* Book 4. The comments on Aristotle's treatment of place and time and Simplicius' own corollaries on place and time have already been translated by J.O. Urmson in two earlier volumes of this series. What remains is the comments on Aristotle's treatment of the void in *Physics* Book 4, chapters 6-9.

Philoponus: On Aristotle *Physics* 5-8

It is of some interest to gain access to a commentary by Philoponus. The first half of his commentary on the *Physics*, Books 1 to 4, which is available in Greek, is full of daring innovation. The 'scientific revolution' of postulating an 'impetus' in dynamics started here before it came to the Latin West, and Philoponus' able reply to Aristotle in defence of motion in a vacuum was to be acknowledged by Galileo.[1] It has been suggested that these and other anti-Aristotelian ideas were inserted by Philoponus in a second edition of the commentary after AD 532, whereas the first edition was produced in 517.[2] Many of the later ideas are tailored to Philoponus' Christian beliefs. The subject matter of Aristotle's work itself might raise our expectations. Book 5 distinguishes changes and types of changes. Book 6 discusses the continuum. Body, space, time and motion are all argued to be infinitely divisible, rather than made up of indivisible units, and the associated paradoxes are brilliantly handled. Book 7

1. Richard Sorabji, 'John Philoponus' in his (ed.) *Philoponus and the Rejection of Aristotelian Science*, London and Ithaca NY 1987, ch. 1.

2. Koenraad Verrycken, 'The development of Philoponus' thought and its chronology', in Richard Sorabji (ed.), *Aristotle Transformed: The Ancient Commentators and Their Influence*, London and Ithaca NY 1990, ch. 11.

introduces the need for a Prime Mover and Book 8 makes the final case for this necessity.

Philoponus' commentary on these books, as summarised in Arabic, is not, however, as striking as his commentary preserved in Greek, on the first four books. But this is no doubt because of the drastic shortening. The commentary on the first four books of Greek will fill six volumes of English translation; the Arabic summary relating to the last four books fills less than one. On the other hand, Paul Lettinck's comparison below of the Arabic with the Greek for the first four books suggests that, where comments are preserved in Arabic, they do not diverge very significantly from the Greek. The most interesting difference, perhaps, is that the Arabic seeks to confirm the theory of impetus by drawing on a phenomenon which Philoponus describes in another commentary (*in DA* 334,40-335,30). Sunlight passing through a coloured glass can throw a pool of colour on a facing stone, and an archer imparts impetus to an arrow, and neither has to impart the effect to the intervening air. The eagerness of the Arabic version to defend the idea of impetus strengthens the view of Fritz Zimmerman that the concept of impetus reached the fourteenth-century West from Philoponus through an Arabic route.[3]

It is also a pity that only two Arabic passages are recorded from Philoponus' commentary on Book 8 of the *Physics*, although both comments reveal the innovative, anti-Aristotelian Philoponus. One passage defends the Christian idea that time began, by insisting that a temporal instant need not bound a preceding as well as a following period. The other defends the Christian idea of bodies being created out of nothing, by pointing out that Aristotle is himself committed to an individual form of whiteness, for instance, being created out of nothing. The example is familiar from Philoponus' other writings.[4]

The Arabic paraphrase presents Philoponus as making independent contributions at other points too. He joins in the discussion begun by Aristotle, continued, as he reports by Alexander, and still of philosophical concern today, of what counts as one and the same change (one and the same event, in Donald Davidson's discussion),[5] or one and the same outcome of change. Philoponus criticises Aristotle and maintains that a person can have the same health or process

3. Fritz Zimmermann, 'Philoponus' impetus theory in the Arabic tradition', in Richard Sorabji (ed.), *Philoponus and the Rejection of Aristotelian Science*, ch. 5.

4. *Against Proclus on the Eternity of the World* 340; 347; 365,3; *Against Aristotle on the Eternity of the World* ap. Simplicius *in Phys.* 1141. Cf. Philoponus *in Phys.* 54,24-5; 55,12-13; 191,9-33.

5. Donald Davidson, 'The Individuation of Events', originally in *Essays in Honor of Carl G. Hempel*, N. Rescher (ed.) Dordrecht 1969, pp. 216-34 and reprinted in *Essays on Actions and Events*, Oxford 1990, pp. 163-80.

of becoming healthy as before, even if it changes in degree, but not if it is altogether interrupted by illness.[6]

Another independent contribution is discussed by Paul Lettinck in the list of interesting passages he picks out in his introduction. It may be summarised as follows. Aristotle's pupil Eudemus had asked how the motion or change undergone by a whole could be the same as the motions or changes undergone by its parts. For the parts of a moving body do not travel only part of the distance, and Philoponus adds that the parts of a whitening body do not go only to a small degree white. Someone, perhaps Eudemus, had said that we do get the result under discussion in the case of growth and diminution. For a person will grow by one cubit when his various members grow by suitable fractions of a cubit. Simplicius was later to discuss whether the time taken by a whole body to make a given journey could be the sum of the motions that the parts would take. He suggests that this would be possible for parts thrown upwards or subjected to motion in other unnatural directions, for then the parts move more easily and in less time than the whole. But with falling bodies, and generally with bodies moving in natural directions, though there will still be a proportionality, it will be the inverse one, because the parts fall *less* easily and in *more* time than the whole.[7]

With this last claim Philoponus would disagree. For he argues in the commentary on *Physics* that even a much heavier body will fall only slightly faster, and the difference in speed will be imperceptible or nothing at all, if the heavier body is only double.[8] He does not contradict this claim when earlier in the same passage he emphasises that a greater weight does indeed fall in (somewhat) less time, nor when in the commentary on *Physics* 4 he argues that two weights of a pound each joined together are (somewhat) more than two pounds.[9]

Since he does not take the same view as Simplicius on falling bodies, he needs a different answer. As Lettinck says, his formula here seems to be that the amount of body moved, when the whole moves, is the sum of the amounts of body moved in the movements of the parts.

Simplicius: On Aristotle on the Void

Aristotle denies the possibility of vacuum or void in *On the Heavens* 1.9, and in *Physics* 4.6-9. A void is thought of as a place deprived of body. But Aristotle does not conceive place like other people, as a three-dimensional expanse (*diastêma*) that goes right through

6. 560,10 ff.
7. Philoponus below 656,10 ff., Simplicius *in Phys. 6*, 473,21-474,8; 977,31-978,16.
8. Philoponus *in Phys*. 678,24-684,10 at 683,17-25
9. Philoponus *in Phys*. 429,7 ff.

things. A person's place is rather a mere contour. It is the inner surface of their immediate surroundings, for example of the air that surrounds them, provided the air is still. That is not the way in which people normally conceive place, when they think of a void as a place deprived of body. The only three-dimensional expanse that Aristotle recognises, other than the body itself, is the volume (*onkos*) of the body. He thinks it would be redundant to postulate yet another kind of three-dimensional expanse running through it, namely a place which might be full or empty.

Believers in the void had thought that it was required to explain how things can have room to move, or contract, or expand, or absorb other things. Aristotle disagrees and provides a series of arguments, which the Stoics sought to counter in detail,[10] to show that motion would, on the contrary, be actually impossible in void.

Aristotle's conception of place had been questioned even by his immediate successor Theophrastus, and it won the allegiance of very few Greeks. Many thought of place as a three-dimensional expanse, and many Platonists had made at least this much concession to the idea of void, that place could exist without body (so far as its own nature was concerned),[11] even if other considerations prevented its doing so.

Simplicius' own view of place, however, was an unusual one, which would not easily permit anyone to conceive of place as empty. Place is not an extension (*diastasis*). My individual (*idios*) place is a dynamic entity that moulds me, arranges my parts in the right order, and accompanies me wherever I go. I also have a place that is shared with others (*koinos*), and which may be an adventitious (*epeisaktos*) place such as *in the market*, or maybe broader still, for example the place of the earth within the cosmic system. That place is not unique to me, but it is to the earth, and it will mould and accompany the earth. Finally, my immediate place within the air which I occupy has no permanent existence, but perishes when I leave it.[12] It is hard to see how any of these places could be conceived as empty or void.

Accordingly, Simplicius agrees with Aristotle's denial of void. But he does not agree with all of his arguments against it and in particular he attacks some of Aristotle's arguments that void would make motion impossible, as Peter Lautner brings out in his Introduction below.

10. Richard Sorabji, *Matter Space and Motion*, London and Ithaca NY 1988, ch. 9.
11. Simplicius *in Phys.* 601,24; 618,24.
12. Richard Sorabji, *Matter, Space and Motion*, 207-11, and *Simplicius: Corollary on Place*, translated in this series by J.O. Urmson.

PHILOPONUS

On Aristotle *Physics* 5-8

FROM THE ARABIC SUMMARY

translated by

Paul Lettinck

Translator's Note

When I had completed the first draft of this translation in the autumn of 1992, Giannakis' thesis *Philoponus in the Arabic tradition of Aristotle's Physics* was published. Part of it is a translation of the same text as is translated here. Giannakis also discusses the origin and structure of the MS containing the text and gives a very useful survey of the contributions of the various Arabic commentators whose comments occur in that MS, and of their relations to Greek commentators.

Moreover, he has compared the complete MS with the edition of Badawî, and has mentioned all instances where this edition should be corrected.

I am very grateful to him for allowing me to see his thesis. I was able to improve my translation at several places, using his corrections of Badawî's edition. I also adopted some of his conjectures where the text was unclear.

Introduction

Aristotle's *Physics* in the Arabic world

The *Physics* of Aristotle was widely studied in the medieval Arabic world. The work was translated into Arabic several times in the ninth and tenth centuries, and was studied and commented upon by Arabic philosophers such as al- Fârâbî, Ibn as-Samḥ, Ibn Sînâ, Ibn Bâjja and Ibn Rušd. The Greek commentaries on the *Physics* of Alexander of Aphrodisias, Themistius and Philoponus were also translated and studied by the Arabs. Simplicius' commentary was not known in the Arabic world.

The history of the translation of the *Physics* and its Greek commentaries into Arabic has been described by Peters.[1] His account is based mainly on the *Fihrist* (988), a list of the books which were known to its author, Ibn an-Nadîm.

The only translation of the *Physics* which has been preserved is that of Isḥâq ibn Ḥunayn (d. 910). It is contained in a MS from Leiden University Library,[2] together with the commentaries of several Arabic authors. This translation with commentaries was edited by Badawî in 1964-65.

Of the other translations of the *Physics* listed by an-Nadîm those of Qusṭâ ibn Luqâ (d. 912) and ad-Dimašqî (d. 900) deserve mention here, because a few quotations from them occur in the Leiden MS. An-Nadîm says that Qusṭâ translated Books 1-4 together with the commentary of Philoponus, and Books 4, 5 and 7 with the commentary of Alexander. He also says that Books 5-8 with the commentary of Philoponus were translated by Ibn Nâ'ima (fl. *c*. 830) and that most of Books 1, 2, 6 and 8 with Alexander's commentary were extant in Arabic.

The commentary of Alexander is not preserved in Greek, although Simplicius' commentary contains many quotations from it. It was, at least for the most part, translated into Arabic, but this translation is not preserved either; a few quotations occur in the above mentioned Leiden MS. Nor is the Arabic translation of Themistius' paraphrase of the *Physics* preserved, but again, a few quotations occur in the Leiden MS. The commentary of Philoponus on Books 1-4 is completely

1. F.E. Peters, *Aristoteles Arabus*, Leiden 1968, 30-4.
2. MS Leiden Or. 583, cf. P. Voorhoeve, *Handlist of Manuscripts in the Library of the University of Leiden and other Collections in the Netherlands*, Leiden 1980², 327.

preserved in Greek, whereas only fragments survive of his commentary on Books 5-8. His complete commentary was translated into Arabic (not preserved), and widely studied. A summary or paraphrase of his commentary on Books 3-7 (and two passages on Book 8) are attached to Isḥâq's translation in the Leiden MS. From this summary one may become acquainted with the contents of some of his lost Greek commentary. It is this summary of which an English translation is presented in this book, insofar as it pertains to Books 5-8. We shall make some general remarks on the summary for Books 3-8 below in this introduction.

The commentaries attached to Isḥâq's translation

The Leiden MS which contains Isḥâq's translation is the outcome of the study of the *Physics* in the Baghdad school of Yaḥyâ ibn 'Adî (d. 973) and his pupil Abû 'Alî Ibn as-Samḥ (d. 1027). Besides the Arabic text of the *Physics* it contains commentaries by Ibn as-Samḥ, Yaḥyâ ibn 'Adî, Yaḥyâ's teacher Abû Bišr Mattâ ibn Yûnus (d. 940), and Abû l-Faraj ibn aṭ-Ṭayyib (d. 1044). In addition, a few comments of Alexander and Themistius are quoted, as well as some phrases from the translations of Qusṭâ and ad-Dimašqî.

Many comments in Books 3-7 are preceded by the name 'Yaḥyâ'; they appear to be a summary or paraphrase of Philoponus' commentary. Thus 'Yaḥyâ' does not refer to Yaḥyâ ibn 'Adî, but to Yaḥyâ an-Naḥwî (John the Grammarian = John Philoponus), as has been suggested by Endress.[3] The comments by Yaḥyâ ibn 'Adî are clearly distinguished from those of Yaḥyâ (an-Naḥwî), as they are preceded by his full name Yaḥyâ ibn 'Adî. That Philoponus' commentary was known and studied in Baghdad at the end of the tenth century is evident from al-Qifṭî, who says that Îsâ ibn 'Alî studied Philoponus' commentary with Yaḥyâ ibn 'Adî.[4]

The editor of the texts in the Leiden MS was Abû l-Ḥusayn al-Baṣrî (d. 1044), a pupil of Ibn as-Samḥ. From the colophon at the end of Book I we learn that he possessed Isḥâq's translation, in a copy belonging to Ibn as-Samḥ, and that he compared this copy with a copy belonging to Yaḥyâ ibn 'Adî, who had compared Isḥâq's autograph with the Greek text three times, and once with the Syriac text.[5] Furthermore we are informed that al-Baṣrî added to Isḥâq's translation commentaries by Ibn as-Samḥ, Abû Bišr Mattâ, and Yaḥyâ. His edition was completed in 1004. Then the texts collected by al-Baṣrî were copied in 1077 by an unknown scribe, and finally by Abû l-Ḥakam al-Maġribî in 1130.

3. G. Endress, *The Works of Yaḥyâ ibn 'Adî*, Wiesbaden 1977, 36ff.
4. al-Qifṭî, *Ta'rîḵ al-ḥukamâ'*, ed. J. Lippert, Leipzig 1903, 245.
5. Aristotle, *aṭ-Ṭabî'a* 76,5-77,15.

Thus, al-Baṣrî studied the *Physics* with Ibn as-Samḥ, and provided it with Ibn as-Samḥ's commentaries. Starting from Book 6.5, however, Ibn as-Samḥ's commentaries cease to occur, and are replaced by those of Abû l-Faraj ibn aṭ-Ṭayyib. Probably al- Baṣrî studied this part with Abû l-Faraj and added Abû l-Faraj's commentaries. This agrees with the fact that there are phrases which in the first part are preceded by 'I said to Abû 'Alî', (i.e. al-Baṣrî asks his teacher a question) and in the second part by 'I said to Abû l-Faraj'. The commentaries of Abû Bišr and Yaḥyâ (the summary of Philoponus) must have been present in the copies of Ibn as-Samḥ, Yaḥyâ ibn 'Adî and Abû l-Faraj used by al-Baṣrî. It is not known who made the summary from Philoponus' commentary. The quotations from Alexander and Themistius as well as the occasional phrases in the translation of Qusṭâ and ad-Dimašqî also would have been laid down by members of the Baghdad school.

Sometimes a commentary is preceded by the names of both Yaḥyâ and Abû 'Alî (Ibn as-Samḥ) or Yaḥyâ and Abû l-Faraj. In these cases al-Baṣrî may have combined two commentaries in one treatise, or possibly Ibn as-Samḥ and Abû l-Faraj were indicating that their commentary was influenced by Philoponus.

A study of the 'Yaḥyâ' commentaries in the Leiden MS and a comparison with what is preserved of the Greek commentary of Philoponus reveals the following:

As has already been mentioned, the comments on Books 3-7 (and two comments on Book 8) which are preceded by 'Yaḥyâ', are mostly a selection from Philoponus' text. Many sentences are a literal translation from Philoponus' Greek text. These comments also follow the way in which Philoponus organises his commentary: first giving a general comment on a section of Aristotle's text, followed by more detailed comments on certain words or phrases.

On the other hand, it appears that the comments of 'Yaḥyâ' are not always just a summary of Philoponus' comments: sometimes a reordering of phrases has taken place, or examples are given which are different from those in Philoponus' Greek commentary. Some instances of this will be given below. On the whole, the Arabic version is rather a paraphrase of the Greek commentary than a mere summary. Such a commentary must have been composed in the Baghdad school of Yaḥyâ ibn 'Adî and Ibn as-Samḥ. From this Arabic paraphrase we become acquainted with some of the lost Greek commentaries of Philoponus on Books 5-8. We shall mention below the most interesting cases.

It further appears that the 'Yaḥyâ' commentaries on Books 3 and 4 use Qusṭâ's translation. Yaḥyâ quotes Aristotle's text numerous times; these quotations are in a version which is not the one of Isḥâq. Sometimes he cites a phrase from Aristotle which is not in Isḥâq's translation at all. At one place in the Leiden MS a phrase is quoted in Quṣṭâ's translation, while this phrase is also quoted in Yaḥyâ's commentary; it appears that Yaḥyâ's quotation is in Quṣṭâ's version, not in Isḥâq's version.[6] The quotations from Aristotle belonging to Yaḥyâ's commentary on Books 5-7 are always given in Isḥâq's version, with a few exceptions in Book 7.[7] All this agrees with the *Fihrist*, according to which Books 1-4 were translated by Quṣṭâ with the commentary of Philoponus, and apparently the Arabic paraphrase was made from this translation. Quṣṭâ did not translate Philoponus' commentary of Books 5-8, and the composer of the paraphrase used Isḥâq's version for the quotations of Aristotle.

A more detailed study of the origin of the Leiden MS and its structure is given by Giannakis.[8] He also gives a valuable survey of the contributions of the various Arabic commentators and their relations to the Greek commentators.

The Arabic paraphrase of Philoponus' commentary

As has been said above, the Arabic commentary of Yaḥyâ is often not strictly a summary, but instead a rather loose paraphrase of Philoponus' commentary: phrases have been reordered, different examples are given, additions have been made, and sometimes even the content of a statement is different. Some examples of the most interesting and characteristic differences will be given here. We shall see that sometimes the Greek text was changed with a view to an Arabic audience. A detailed comparison between the passages of Philoponus and Yaḥyâ is given by Lettinck and Giannakis.[9] Moreover we shall discuss here the commentaries on Books 5-8 of which no corresponding Greek fragment is preserved. The most interesting cases are an argument against infinite time (a comment on 5.2 226a1ff.[10]), a comment on the divisibility of changing bodies (on 6.4, 234b10ff.[11]), a comment on the divisibility of motion according to a division of the moving body (on 6.4, 234b21ff.[12]), a comment on Zeno's paradox of the stadium (6.9, 239b32ff.[13]) and another argument against infinite time (a comment on 8.1, 251b10[14]). Before discussing these cases, I shall first illustrate the kinds of way in which the Arabic version differs, where we have the Greek to compare with it.

6. See Lettinck, *Aristotle's Physics* 293-4.
7. See below the translation of Book 7, n. 252.
8. Giannakis, *Philoponus*, part one.
9. Lettinck, *Aristotle's Physics*; Giannakis, *Philoponus* 152-82.
10. Yaḥyâ in Aristotle, *aṭ-Ṭabî'a* 523,12 ff.
11. Yaḥyâ in Aristotle, *aṭ-Ṭabî'a* 649,24 ff.
12. Yaḥyâ in Aristotle, *aṭ-Ṭabî'a* 656,10 ff.
13. Yaḥyâ in Aristotle, *aṭ-Ṭabî'a* 722,13 ff.
14. Yaḥyâ in Aristotle, *aṭ-Ṭabî'a* 816,14 ff.

In Book 3, Chapter 1 Aristotle discusses the concept of motion or change (*kinêsis*). He remarks that change is always change in respect to one of the categories of substance, quantity, quality or place. Thus there are four kinds of change and they do not have a common genus (category) to which they all belong (200b34). In his comment on this passage Philoponus compares motion to the knowable and says that the knowable also belongs to different categories (for substance, quantity etc. are knowable), but as knowable it belongs to quality.[15] In the Arabic summary, however, it is said that the knowable belongs to the category of the relative.[16]

In Chapter 3 of the same book Aristotle claims that motion may be considered under two aspects, as an action of the mover on what is moved and as an effect suffered (being-acted-upon) by what is moved from the mover, but that these are two accounts of one and the same process. He compares this to the road from Thebes to Athens and the road from Athens to Thebes: they are in fact the same, but differ in account (202b13), for the one goes uphill and the other downhill. Philoponus quotes this example.[17] In the Arabic version it is said that the road between Basra and Baghdad is one and the same thing, but that going from Basra to Baghdad is not the same in definition as going from Baghdad to Basra.[18] The translation of Isḥâq has kept the names of the Greek cities.

In Chapter 5 of Book 3 it is shown that it is impossible for an infinite physical body to exist. One of the arguments is that a body must be light, heavy, or partly light and partly heavy. It cannot be infinite and light, for then it would have to be in the place 'above', so that this place would be infinite (and no room would be left for 'below'); similarly, it cannot be infinite and heavy (205b24ff.). In the Arabic paraphrase of Philoponus' commentary the following argument is given for the impossibility of the existence of an infinite body which is partly light and partly heavy:[19] If such a body existed, then either both parts would be infinite, or one of them would be infinite and the other finite. In the first case this would mean that the infinite is doubled, in the second case that it is increased. This argument is not in the Greek commentary of Philoponus on this passage, although it agrees with his commentary on other passages in which infinite bodies or numbers are discussed.[20]

15. Philoponus, *in Phys.* 349,8-11.
16. Yaḥyâ in Aristotle, *aṭ-Ṭabî'a* 175,26.
17. Philoponus, *in Phys.* 373,19-21 and 374,5-11.
18. Yaḥyâ in Aristotle, *aṭ-Ṭabî'a* 198,1-2.
19. Yaḥyâ in Aristotle, *aṭ-Ṭabî'a* 249,3-5.
20. In his comments on *Physics* 3.5 many of Philoponus' arguments against the actual infinite may be summarised by saying that it is impossible to increase or to multiply the infinite; i.e. he considered the statements $\infty + c = \infty$ and $c \times \infty = \infty$ to be contradictory. Another formulation is that two infinites cannot be unequal. This kind of argument became widespread in the medieval Arab and Latin world.

In Book 4, Chapter 1 Aristotle gives arguments for and against the existence of place. An argument for the existence of place is the fact of mutual replacement: when water goes out of a vessel it is replaced by air, so what first contains water and then air is something existing, where bodies go in and out, and which is different from these bodies themselves (208b1ff.). In the Arabic paraphrase of Philoponus' commentary it is said that if one argues that from the fact that bodies replace one another it follows that there must be something in which this replacement occurs, then the fact of replacement does not contribute to the argument.[21] For if a body is not replaced, we also assume that it is in something. If we did not assume this, we could not assume that replacement occurs in something. This comment is not in the Greek text of Philoponus.

Further additions in the Arabic paraphrase, which are not in the Greek text, occur in the commentary on Chapter 7 of Book 4. Here some arguments are refuted which had been adduced to prove that the existence of a void is necessary to explain certain phenomena, like motion and increase in volume (214a26ff.). In the Arabic version, a rotating millstone is given as an example of a body which does not require a void for its motion.[22] A second example involves the increase in volume that occurs when water changes into air: this does not mean that two bodies coincide, nor that a void is required to accommodate the increase. When air arises from water the surrounding air may be compressed, or it may partly change into water, which implies a decrease in volume. In both cases space becomes available for the air which came into being out of water.[23] The suggestion that the surrounding air may be compressed is not in the Greek commentary.

An argument against the existence of the void brought forward by Aristotle in Book 4, Chapter 8 is that a body which is thrown would not be able to move in a void after its contact with the thrower has been broken, because there would be nothing to move it. Aristotle's view is that it is the air which moves the body in one of two ways (215a14ff.). An account of this view is given in the Arabic paraphrase as follows: The motion of an arrow after it has been shot by a bow occurs because the air in front of the arrow moves to the back and pushes the arrow, or because the string of the bow pushes the arrow and the air, and this air propels the arrow further, until it reaches its highest point. Then the force which the air has received from the thrower becomes weaker, and the arrow falls. The force of the air at the beginning of the motion is stronger than the tendency of the arrow to move downwards.[24] The Greek commentary does not mention the

21. Yaḥyâ in Aristotle, *aṭ-Ṭabî'a* 276,8-9.
22. Yaḥyâ in Aristotle, *aṭ-Ṭabî'a* 354,11-12.
23. Yaḥyâ in Aristotle, *aṭ-Ṭabî'a* 355,8-11.
24. Yaḥyâ in Aristotle, *aṭ-Ṭabî'a* 369,23-5.

arrow, nor does it say what happens after the arrow has reached its highest point.

Philoponus does not agree with Aristotle's view on the motion of thrown bodies. He brings forward his own view, which is that thrown bodies are not propelled by the air, but are moved by an internal, incorporeal force (impetus) which is impressed into them by the thrower. As this force grows weaker the motion slows down and eventually ceases. In the Arabic paraphrase the explanation of this view is concluded by the following remark, which does not occur in the corresponding Greek passage: The concept of an incorporeal force which is transmitted from the thrower to the thrown body is not strange, because the same thing happens in the process of seeing: a visible body impresses something in the body of our eye. Something similar occurs with the light of the sun. It may get coloured by passing through a coloured piece of glass, and when this light falls on a stone it gets this same colour, so the colour is transmitted to the stone.[25]

Another argument against the existence of the void from Book 4, Chapter 8 starts from the assumption that the speed of a body moving through a medium is inversely proportional to the density of the medium (215a22ff.). If a body could move through a void, then the proportion of its speed to its speed in a plenum would be the inverse of the proportion of the (density) of void to the (density) of the plenum. But the void has no proportion to a plenum, just as nothing (zero) has no proportion to a number such as four. Four does have a proportion to smaller numbers like three, two and one, which are such that one may say that four is composed of one of these smaller numbers and another number (e.g. $4 = 3 + 1$). But one cannot say that four is composed of nothing (zero) and four, so four has no proportion to nothing.

Philoponus gives an account of this argument, and, in the Greek text, adds: 'In the same way a plenum has no proportion to the void, for it would be ridiculous to say that *water* is composed of void and water.'[26] The corresponding phrase in the Arabic version says that it would be nonsense to maintain that *air* is composed of void and air.[27]

Aristotle gives another argument against the existence of a void, namely that bodies of different weights would fall equally fast in a void, because there is no plenum which must be pierced (216a11ff.). Philoponus gives as examples of such bodies a body of lead and one of cork (Greek text[28]). The Arabic paraphrase gives as example a stone of a fruit and a feather.[29]

25. Yaḥyâ in Aristotle, *aṭ-Ṭabî'a* 371,21-4.
26. Philoponus, *in Phys.* 648,28-9.
27. Yaḥyâ in Aristotle, *aṭ-Ṭabî'a* 374,17-18.
28. Philoponus, *in Phys.* 661,4.
29. Yaḥyâ in Aristotle, *aṭ-Ṭabî'a* 381,21-2.

In 216a26 ff. Aristotle compares the consequences of placing a *wooden* cube in water (or air) and in a void. The Greek text of Philoponus[30] adopts Aristotle's example, but the Arabic version uses an *iron* cube.[31]

In his *Corollary on Void* Philoponus criticises Aristotle's arguments from Book 4, Ch.8 that motion through a void would be impossible because, in the absence of resistance, it would (absurdly) take no time.[32] Philoponus replies that all motion takes time. What the lack of resistance obviates is the need for *extra time*. The corresponding Arabic version has additions which are not in the Greek text. For instance, it is said that the time needed to cover a distance in a void is less than in a plenum (though never zero), but not because a void and a plenum both resist the motion and a void resists it to a lesser extent than a plenum; for then the void and the plenum would have a proportion to each other.[33]

In Chapter 12 of Book 4 several aspects of time are discussed. Time is a measure of motion, so it will indirectly be a measure of rest too (221b7ff.); i.e. it measures one thing's rest by means of the motion of something else which occurs during that rest. The Arabic version of Philoponus' commentary contains an illustration which is not in the Greek text: Something may come to rest at the beginning of the day and remain at rest during all that day, until the end, and then finish its rest. Then we say that time measures this rest by means of the motion of something else; for both rest and motion have the same time.[34]

In Chapter 14 of Book 4 Aristotle says that as time is a measure of change this implies that the change is counted (measured). If there were no soul no counting could take place, and no time would exist. There are some differences between the Greek commentary of Philoponus on this subject and its Arabic paraphrase. The Greek text says that when there are ten stones they exist as a number (ten), even if there existed no soul (to do the counting).[35] The Arabic version has ten horses instead of ten stones.[36] Later on the Greek text reads: 'One and the same number in the soul, like ten, counts all collections of ten which exist outside the soul, such as horses, men, stones, etc.'[37] The Arabic version gives as examples: horses and camels.[38]

30. Philoponus, *in Phys.* 661,17.
31. Yaḥyâ in Aristotle, *aṭ-Ṭabîʿa* 382,2.
32. Philoponus, *in Phys.* 667,8-675,11.
33. Yaḥyâ in Aristotle, *aṭ-Ṭabîʿa* 393,8-13.
34. Yaḥyâ in Aristotle, *aṭ-Ṭabîʿa* 459,16-18.
35. Philoponus, *in Phys.* 770,3.
36. Yaḥyâ in Aristotle, *aṭ-Ṭabîʿa* 475,5.
37. Philoponus, *in Phys.* 776,23-4.
38. Yaḥyâ in Aristotle, *aṭ-Ṭabîʿa* 482,20.

I now come to the books for which the Greek is only fragmentary, and will pick out some of the more characteristic and more interesting aspects of the Arabic version. In Book 5, Chapter 2 Aristotle seeks to show by several arguments that there is no change of change, irrespective of the sense in which 'change of change' is taken. For instance, the transition from rest to being in change is not a change. For otherwise every change would need an infinite number of preceding changes; this infinite series would not have a first member, so the assumed change would not exist either (225b33ff.).

Philoponus gives the following comment in the Arabic version (there is no corresponding passage in the surviving Greek fragments): 'By means of this argument one may show that Socrates cannot have an infinite number of ancestors. For if the number of his ancestors were infinite, there would not be a first ancestor, so the generations following that ancestor, including Socrates, would not exist either.'[39]

The conclusion, which is not explicitly stated in this comment, is that time is not infinite, but has a beginning. This is a typical Philoponian argument against the infinity of time. He presents it also in other works.[40] In fact the argument is not offered in support of Aristotle's proposition that there cannot be change of change or generation of generation, but as a refutation of Aristotle's statement from *Physics* 8.1 that every change is preceded by another change, which implies that time is eternal. Philoponus and those who criticise his argument consider both statements (from *Physics* 5.2 and 8.1) to be about infinite causal chains.

Philoponus' argument is criticised by adherents of the eternity of the world, such as Ibn Bâjja and Ibn Rušd, who distinguish between accidental and essential infinite causal chains. In an essential chain of changes the existence of a former member is necessary for the existence of the next one, because the next one is essentially caused by the former. In an accidental chain this is not the case; there the changes are essentially caused by something else. They consider Philoponus' argument to be valid for an essential infinite causal chain of changes, and this is believed to be impossible indeed, but if the succession is accidental, an infinite series is not impossible, as long as an infinite period of time is available. The generation of a son by his father is such an accidental succession; a man is essentially generated by the First Mover (or: Active Intellect). Thus, if time is eternal it is not impossible for Socrates to have an infinite number of

39. Yaḥyâ in Aristotle, *aṭ-Ṭabî'a* 523,12.

40. See R. Sorabji, *Time, Creation and the Continuum*, London and Ithaca NY 1983, 228-9; H.A. Davidson, *Proofs for the Eternity, Creation and the Existence of God in Medieval Islamic and Jewish Philosophy*, New York and Oxford 1987, 87-8; H.A. Wolfson, *The Philosophy of the Kalâm*, Cambridge (Mass.) and London 1976, 413-14.

ancestors.[41] Philoponus provides another argument against infinite time at *in Phys.* 428,14-429,20.

In Book 5, Chapter 4 Aristotle states the conditions under which two motions may be called one and the same. Two motions are numerically the same if the following things are the same: the subjects of the motion, the path, goal and manner of motion, and the time (i.e. if the motions are simultaneous and of equal duration). The subject must be the same essentially, not only accidentally (227b31ff.). This last phrase is illustrated by an example which is adopted by Philoponus in his commentary. This commentary runs as follows in the Greek text: If Coriscus is walking and becomes black in the same time, then these motions are not the same, because their subjects are the same only accidentally, not essentially. The subject of walking is Coriscus and the subject of becoming black is the white; it is accidental that Coriscus is white.[42] The Arabic version has: If Socrates is learning and becoming hot at the same time, then these motions are not the same; they are not even one by genus, because although the motions belong to the same genus, sc. qualitative motion, and the time is the same, they do not both belong to the same subject essentially. The subject of learning is indeed Socrates, but the subject of becoming hot is what is cold, and it is accidental for Socrates that he is cold.[43] Yaḥyâ continues (there is no corresponding passage in the Greek fragments): The example of someone learning and becoming hot is more suitable than the example Aristotle uses of someone walking and becoming black, because in the latter case the motions themselves are already different in genus (local motion and qualitative motion), and this fact is sufficient to show that they are not one; the argument that the subjects are not the same essentially is not necessary. In the former example the motions are the same in genus, thus in order to show that they are not one, the fact that the subjects are different is needed.

The subject of Book 6 is the continuity and infinite divisibility of magnitudes, motions and time. The paradoxes of Zeno are discussed as part of this subject. These paradoxes are arguments which attempt to show that motion along a finite distance is impossible in a finite time. One of these arguments is that if a body were to traverse a finite

41. See Ibn Bâjja, *Commentary on the Physics* 51,14-54,2 (on *Physics* 5.2) and 174,21-175,18 (on *Physics* 8.1); Ibn Rušd, *Long Commentary* 218B8-K4 (on *Physics* 5.2), 349G14-350M1 (on *Physics* 8.1, with a reference to *Physics* 5) and 388K8-L4 (on *Physics* 8.6); Ibn Rušd, *Short Commentary* 41,19-42,4 (on *Physics* 3), 76,3-77,4 (on *Physics* 5.2) and 133,5-135,11 (on *Physics* 8.1, with a reference to *Physics* 5); see also the works mentioned in the preceding note: Sorabji 230, Davidson 131-3, Wolfson 428-30 and further Lettinck, *Aristotle's Physics* 658-60. This last reference gives an account in English of the above-mentioned Arabic and Latin texts.

42. Philoponus, *in Phys.* 856,11-15 and 794,2-5.

43. Yaḥyâ in Aristotle, *aṭ-Ṭabîʿa* 557,15 ff.

line in a finite time, it would be in contact with an infinite number of things (i.e. the half-way point, the three-quarters-way point and an infinity of other points on the line) in a finite time, and this is impossible. Aristotle explains why this argument is mistaken and then proves that for a body to traverse a finite distance a finite time is sufficient. Suppose the finite distance AB were traversed in an infinite time t; then consider a finite part TU of t. In TU a part AC of AB is traversed. AC will be an exact measure of AB, or less (*elleipsei*) or more (*huperbalei*) (i.e. a multiple of AC is equal to AB, or it is smaller or greater than AB). Then AB will be covered in a time which is a multiple of TU, or a finite time more or less, and that is finite (6.2, 233a34ff.).

In the Greek text Philoponus gives the following examples of the relation between AB and AC: If AC is one cubit and AB is seven cubits, then AC is an exact measure of AB, for seven times AC equals AB. If AC is two cubits and and AB is seven cubits, then four times AC exceeds AB, and three times AC is less than AB. In any case, there is always a finite proportion between AC and AB.[44] The Arabic version says that if AC is two cubits and AB is seven cubits, then three times AC results in a length which is one cubit shorter than AB, and this is less than AC, whereas two times AC gives a length which is three cubits less than AB, and that is more than AC.[45] Apparently the Greek text was not correctly understood.

In Chapter 4 of Book 6 Aristotle first shows that what is changeable must be divisible: if a thing is undergoing a change it must be partly in the state at which the change starts, and partly in the next stage of the change (234b10 ff.). No comment by Philoponus on this passage has been preserved in Greek. The Arabic version gives a combined comment by Yaḥyâ and Abû 'Alî ibn as-Samḥ. After an account of Aristotle's argument a problem raised by Alexander is brought forward,[46] namely that Aristotle's argument is valid only for things which gradually change, one part after another, not for instantaneous changes, such as the coagulation of milk or the darkening of a face exposed to the sun. Alexander solved the problem, according to this comment, by noting that Aristotle here discusses only non-instantaneous change, such as the burning of wood. It is not necessary to treat of instantaneously changing things, because it is evident that they are divisible, since 'instantaneous change' means a change which occurs simultaneously in all *parts* of a body.

This question raised by Alexander has also been taken up by

44. Philoponus, *in Phys*. 803,14-19.
45. Yaḥyâ in Aristotle, *aṭ-Ṭabî'a* 632,27 ff.
46. Yaḥyâ in Aristotle, *aṭ-Ṭabî'a* 649,24 ff. See further Sorabji, op. cit., 410-11.

others: Theophrastus, Themistius,[47] Simplicius,[48] Ibn Bâjja[49] and Ibn Rušd.[50] The report of Alexander's solution by Yaḥyâ (Philoponus) contradicts what Themistius, Simplicius, and Ibn Rušd say about him: they claim that according to Alexander there are no instantaneous changes; changes which seem to be instantaneous in fact occur in an imperceptible time. What Philoponus reports here is in fact Themistius' solution. On the other hand, it is remarkable that Alexander, in his commentary on Aristotle's *De Sensu*, agrees with Aristotle that the freezing of water may be instantaneous and that the illumination of an area by light is always instantaneous.[51] Thus the ancient commentators were not united in their interpretation of Alexander's opinion.

Themistius discusses the problem as follows: When something changes instantaneously, how can one say that it is partly white and partly black?[52] This question had already been asked by Theophrastus, and dealt with by many other commentators. Some said that all parts of a body never change in one moment at the same time, but that a part always changes first, then another one, e.g. when milk is curdled, or a body becomes dark when it is exposed to the sun. For there is always a part which is most liable to change or which is closest to the cause of the change, and this part will change first.[53] This is not true, however: we observe that things as a whole change instantaneously and Aristotle himself mentions it in Book 1 (186a15). Our opinion is that it is not necessary for Aristotle to prove that instantaneously changing things are divisible, because it is evident that this is the case. For if we say that a body changes instantaneously, we mean that it changes at one moment in all its parts. So the body must have parts.

In his paraphrase of *Physics* 6.6 Themistius discusses instantaneous changes again in connection with the proposition that what has changed has arrived at its new state after a process of change and that this does not apply to instantaneous changes.[54] He says that Alexander's opinion was that all changes take some amount of time, and that Theophrastus has raised the problem of instantaneous change by considering the change from shadow to light: when a lamp is brought into a room, the room becomes illuminated instantane-

47. Themistius *in Phys.* 191,30-192,22 and 197,1-19.

48. Simplicius *in Phys.* 966,15-969,24 and 997,30-998,19.

49. Ibn Bâjja, *Commentary on the Physics* 96,7-104,8.

50. Ibn Rušd, *Long Commentary* 265I12-266F12; id. *Short Commentary* 99,2-101,9.

51. Alexander, *in De Sensu*, P. Wendland (ed.), *CAG* III1, Berlin 1901, 131,21-135,22.

52. Themistius *in Phys.* 191,30-192,22.

53. Alexander's opinion, as appears from Simplicius' report and from another comment of Themistius, on 6.6, see below.

54. Themistius *in Phys.* 197,1-19.

ously, not one part after another. But this case is, according to Themistius, similar to that of 'making contact' (*haphê*), which indeed occurs instantaneously. This is not a process of change however, but only a state of being which exists without having come to be, i.e. without having been preceded by a change.

Simplicius gives an extensive discussion of the question of the divisibility of instantaneously changing bodies in which, also, the opinions of Alexander and Themistius are reviewed.

After having shown that everything that changes is divisible, Aristotle argues that the change (motion) itself is also divisible. It is divisible in two ways: according to the divisibility of time and according to the divisibility of the moving object. Thus there exists a corresponding divisibility of time, motion, the moving body and the path along which the motion takes place.

Philoponus realises that the statement that a motion is divisible according to the divisibility of the moving object is problematic. For what does it mean, e.g. for a local motion, that the motion may be divided into parts which are the motions of the parts of the moving body, or in a different formulation that the motion of the whole is the sum of the motions of the parts? It cannot mean that if a body which is divided into ten parts moves a hundred cubits, each part moves ten cubits. Thus, the parts of the motion cannot be motions along parts of the distance, for all parts of the body move the same distance and the body as a whole also moves that same distance, not the sum of the distances covered by each part. This question is extensively discussed by Philoponus (not in the preserved Greek fragments).[55] His solution is that indeed every part of a moving body covers the same distance as the whole, and it is nonsense to say that one should add the distances covered by each part to get the distance covered by the whole. Nevertheless, one can say that each part has its own particular motion, different from that of any other, and different from the motion of the whole, and that the motion of the whole is the 'sum' of the motions of the parts. Apparently the differences among the motions of each part are not due to a difference in the distance covered, but to the difference in the subject of the motion (in accordance with 5.4). One may account for this interpretation if one assumes that 'motion' equals mass or volume times distance. Then the statement that the 'motion' of the whole is the sum of the 'motions' of the parts is correct. The question is different for a change like growth. If each part of a body grows a certain amount, then the total growth is the sum of the partial growths.

Simplicius also discusses the problem, asserting that it had al-

55. Yaḥyâ in Aristotle, *aṭ-Ṭabī'a* 656,10 ff.

ready been recognised by Eudemus.[56] He further remarks that in the case of a forced motion one may say that a part of a body, if separated, would cover a distance in a time which is a part of the time in which the whole covers that distance, in such a way that there is the same proportionality between the times as between part and whole. In the case of natural motion one might maintain the same proposition, but then the proportionality between the times must be taken as the inverse of the proportionality between part and whole.[57] It is clear that such divisions cannot be what is meant when one says that the motion of a body is divisible into the motions of each of its parts.

Chapter 9 of Book 6 is almost completely devoted to a discussion of Zeno's paradoxes, one of which is the problem of the stadium. Aristotle's text presents several difficulties and modern commentators have not arrived at a unanimous interpretation. Philoponus comments on this portion of Aristotle's text are not among the fragments which have survived in Greek. The Arabic version gives a clear explanation of the problem.[58] It appears that this explanation is similar to the one given by Ross,[59] although there is some difference in formulation.

In Book 7, Chapter 2 Aristotle explains that mover and moved must be in contact. This is clear for local motion, as local motion occurs either by pulling or pushing: what pulls or pushes moves together with what is pulled or pushed. There also seems to be another kind of pulling: wood pulls (attracts) fire towards itself without being in motion (244a12ff.). In his commentary on this phrase, which is not preserved in the Greek fragments, Philoponus criticises this example, and presents a better one: a magnet which attracts iron, and amber which attracts dust particles.[60] These are attractors which are not in motion together with what they attract. It seems that these attractors are not in contact with what they attract. But the magnet attracts iron because particles are issued from the magnet which penetrate the iron, so that it is attracted. Another explanation is that the air between magnet and iron is changed so that it gets an attracting power. On either account, the iron is in contact with what attracts it. Similar comments are given by Simplicius,[61] Ibn Bâjja[62] and Ibn Rušd.[63]

The purpose of Book 8 is to prove that there is a First Mover which

56. Simplicius *in Phys.* 973,21-974,8.
57. Simplicius *in Phys.* 977,31-978,16.
58. Yaḥyâ in Aristotle, *aṭ-Ṭabî'a* 722,13 ff.
59. Ross, *Aristotle's Physics* 660ff.
60. Yaḥyâ in Aristotle, *aṭ-Ṭabî'a* 755,21 ff.
61. Simplicius *in Phys.* 1055,6-10.
62. Ibn Bâjja, *Commentary on the Physics* 119,15-122,14.
63. Ibn Rušd, *Long Commentary* 315C5-F6; id. *Short Commentary* 119,3-120,13.

is the ultimate cause of all motions, and which is eternal, incorporeal and unmoved. In Chapter 1 Aristotle shows that it is impossible to maintain that motion at first did not exist and then came into being because every motion which has a beginning is preceded by another motion. Thus, motion has always existed. This is already clear from the fact that time is eternal. For time is a measure of motion, so if time is eternal, motion must be eternal too. Time must be eternal, because if it had a limit (beginning or end) the limit is an instant, and every instant is the beginning of future time and the end of preceding time (251b10ff.).

The Arabic text contains an objection by Philoponus to this statement which is not preserved in the Greek fragments.[64] He says that it is not necessary for each instant to be the beginning of the preceding time and the end of the following time. If one asserts that outside the heavens there is nothing, neither a void nor a plenum as Aristotle does then it is not unreasonable to assume that time is generated, i.e. that there was no time before its generation.

Philoponus' opinion that time is not eternal is well-known from other passages, e.g. *in Phys.* 762,2-9. There he says that someone who does not maintain that time is eternal will not admit that any instant is the end of preceding time and the beginning of following time. He also refers to his commentary on Book 8, saying that Aristotle attempts to prove the eternity of time by means of the eternity of motion, but that he (Philoponus) will show in his commentary on Book 8 that he proves anything but that.

*

The most interesting points and characteristic differences between the Greek text of Philoponus and the Arabic paraphrase have been reviewed above. Other differences will be mentioned in the notes of the translation. If a passage from the Arabic version has a corresponding passage from the Greek text of Philoponus, this will be mentioned in the notes.

Deviations from the Arabic text will be mentioned in the notes. Words which should be deleted are marked off by asterisks. Words which should be added are within angle brackets < >. Words within square brackets [ ] are not emendations of the text, but additions which clarify the meaning. Whenever we mention the Arabic text we mean the text as it exists in the Leiden MS. The edition of Badawî does not always agree with it; deviations in his edition from the MS are also mentioned in the notes. In order that the commentary can be read without Aristotle's text at hand, summaries of relevant

64. Yaḥyâ in Aristotle, *aṭ-Ṭabî'a* 816,14 ff.

portions of the *Physics* have been provided. These are indicated by curly brackets { }.

Finally, a remark about the translation of the terms for 'change' and 'motion': From Book 5 onwards Aristotle makes a clear distinction between 'change' (*metabolê* – *taġayyur*) and 'motion' (*kinêsis* – *ḥaraka*). There are four kinds of change: generation and corruption (change of substance), growth and diminution (change of size), alteration (change of quality) and local motion (change of place). 'Motion' comprises only the three last kinds of change. This means that 'motion' is not only used for local motion, but also for changes such as growth and change of attributes (e.g. change of colour). Such use of the English word 'motion', however, can result in an awkward translation which may have misleading connotations. On the whole, I have translated *taġayyur* (*metabolê*) as 'change' and *ḥaraka* (*kinêsis*) as 'motion', but I have tried to catch the places where this makes particularly odd English and in those places I have rendered the Arabic counterpart of *kinêsis* as change[k].

PHILOPONUS

On Aristotle *Physics* 5-8

Translation

Philoponus: On Aristotle *Physics* 5-8 from the Arabic Summary

[Aristotle 5.1]

{224a21-34: A thing may be a cause or subject of change in three ways: accidentally, by virtue of one of its parts or in itself.}

Yaḥyâ and Abû 'Alî:
In the last four books he [Aristotle] discusses motion and explains in 490,8
which categories motion occurs and in which ones it does not occur.
Before that he explains the difference between change and motion; 10
he says that every motion is a change, but not every change is a
motion, because generation is a change, but not a motion. First he
presents an account which will be used in [explaining] the difference
between motion and change, namely an account of the different
classes of moving things.

Yaḥyâ:[1]
We may divide things into classes in different ways. For instance, we
may divide animals into classes derived from their essences and say 15
that there are rational and non-rational animals or that there are
mortal and immortal animals. We may also divide them into classes
derived from their place of origin and say that there are land-animals
and sea-animals. Furthermore, we may divide them into classes
derived from their limbs and say that there are animals with legs and
without legs. In the same way moving things may also be divided in
different ways. We may divide them into classes derived from the 20
essence of their motion and say that there is local and non-local
motion. Non-local motion may be divided into alteration and growth
and diminution. What moves may also be divided according to its
relation to the motion, i.e. into what is more entitled to be described 491,1
as moving and what is less entitled to be so described; then we say
that what moves does so either by itself (essentially), or accidentally,
or partially, by virtue of a part. Each class of the first division of
motion, i.e. local motion, alteration and growth and diminution, may
be divided into these three classes: what moves by itself, accidentally 5
and partially.

What moves locally may be moving accidentally, for instance when one says, 'The white is walking and is changing its place', for the subject [of the white] is walking by virtue of being an animal, not of being white, and it is accidental for the subject that he is white *by virtue of the fact that the white is walking*. <What moves partially> moves by virtue of something else, namely a part of it which is moving. <An example>[2] is the statement: 'The sleeper is moving or has moved', when his hand moves.

What moves essentially is what has the motion in itself, not by virtue of a subject, nor of one of its parts, e.g. when one says: 'A man is walking', for walking belongs to him not by virtue of a subject, nor of a part, but he has walked and moved in his entirety.

Things which cause motion (movers) may also be classified in these ways; we say that Zayd moves a stone when in fact his hand moves it. When we say: 'The white causes motion', then that is [causing motion] accidentally, because it is accidental for the mover that he is white. When we say: 'A man is moving and walking', then walking belongs to him and he is the effective cause of it in his entirety, not by a part, nor did he become the effective cause by a subject being the cause.

These classes also apply to alteration. For instance, something alters by accident when we say: 'The white has become hot', for if some subject becomes hot, and this subject is accidentally white, then one may say that the white has become hot. An example of something which alters partially is our statement: 'The man has recovered', or 'has become hot', when his hand has become hot or his eye has recovered. The whole has been described as moving by virtue of the motion of a part of the whole. An example of what alters by itself is our statement: 'The water has become hot or cold.'

These classes also apply to growth and diminution. An example of
492,1 accidental [growth] is the statement: 'The white has grown', and of partial [growth]: 'Zayd has grown', when only one of his limbs has grown because only the matter of that limb was submitted to growth. It is possible, then, for someone to have some big limbs, while others are small, and the limbs which become big and those which remain small do so without relation to the body as a whole. We also say: 'Zayd has become smaller and has shrunk', when one of his limbs has shrunk. Similarly, when someone has grown or become smaller by himself one says: 'This person has grown', when he has increased in three dimensions; one says: 'He has shrunk', when he has become smaller in three dimensions.

These three classes also apply to substantial change. We say [as an example] for substantial change by itself: 'Air has become fire.' We say [as an example] of accidental change: 'The white has become water', when the subject of the change was white accidentally. We say [as an example] of partial substantial change: 'The rug has turned into sashes', when the red part of it has turned into sashes.

{224a34-224b6: Motion occurs in that which is moved, not in the mover, nor in that toward which or from which the change proceeds.}

Yaḥyâ and Abû 'Alî:
It is his intention in this passage to explain what [kind of] thing 493,7
motion occurs in, for this will be used in a later lesson. It is clear that when motion occurs there are five things: what is in motion, what causes the motion, the time, that from which the motion starts and 10
that to which the motion proceeds. For instance, for local motion to occur there must be a mover, e.g. the soul, and something which is moved, e.g. the body. Further there must be a time in which the motion occurs, a place from which the motion starts and a place where the motion ends.

If the motion were to be in the mover[3] then it would be necessary that it is moving, because it is not reasonable that something in which 15
the motion exists should not be moving. If it is moving then it must have a mover. The same reasoning applies to the motion of this mover, and this goes on without limit. This has been elaborately discussed in the third book of this work.[4] If the motion were to be in time, since time is the measure of motion, then the motion would be in the measure of motion, which is absurd. If the motion were to be in that 20
from which it starts or that to which it proceeds, then motion would be rest, because motion is a proceeding towards a state of rest, at 494,1
which it comes to an end. If the point at which the motion ends were to be motion, although the point where it ends is the [point for the] stopping and ceasing of the change, then stopping would be motion. A similar argument holds if the motion were to exist in that from which it starts. Our expression 'that from which the motion starts' means: the point where the motion begins; and before the beginning there is no change. Thus what is called 'the beginning of motion' is 5
rest, namely the [state of] rest from which the motion starts, because if that were not rest and standing still, but change and motion, the statement that the beginning of motion is that from which it starts would be invalid.

Let us take local motion as an example. When a body moves from one place to another and the beginning of its motion is, for instance, from below, then necessarily its being below is not a motion, for otherwise, if it were a motion and a change, it would not be the 10
beginning of the motion starting from there. This means that motion arises only when the body leaves that state. Also, if the motion were to exist in the goal towards which the moving body proceeds – we

mean by our expression 'the goal towards which it proceeds' that where the motion stops and where there is no change anymore – then rest and stopping of motion would be motion, unless someone would like to call stopping and halting a motion. But then he would give to 'motion' a different meaning from that which we have posited to be the meaning of motion. To sum up, the beginning of a motion is a standing still, and the same holds for the end of a motion. Motion means change and passing on, and these are different from stopping and halting.

23 **224b2** For example, wood, hot and cold are three different things.

Yaḥyâ:
He presents an illustration of the statement that the moving subject
495,1 is not the same as the form towards which the motion proceeds, nor as the form from which the motion starts. The illustration offered is the wood which is moving from coldness to heat; the wood is different from the coldness from which the motion starts, and different from the heat where the motion ends.

224b5 For form does not cause motion, nor experience motion, and the same is true for place and magnitude.

Yaḥyâ:
He wants to inform us about that in which motion occurs, with respect to all kinds of motion: local motion, growth and alteration. Motion always ends either at a form, or a place – this is local motion – or a magnitude.

Yaḥyâ:[5]
496,5 [224b6-13] He has not explained that motion does not occur in the time, in the sense that the time [in which a motion occurs] would be the subject of the motion, because that is so obvious as to be not worth mentioning. For time is one of the things which exist outside the mover and what is moved. He has not refuted the claim that motion occurs in that from which the motion starts either, because nobody would think that. One might wonder, however, whether motion occurs in the goal towards which the motion proceeds. He refutes the existence of motion in the goal [by arguing] that everything which moves, moves on account of something which it desires to reach as the completion of its form. The absurd and unreasonable conclusion would be that something on account of which the moving subject is moving, namely the goal towards which it moves, would be in motion itself. Aristotle says that the name of the motion is not taken from

its starting-point, but rather from its goal. For we say that what is proceeding towards its perishing is perishing, because the end-point is the perishing, and we do not say it is coming into existence or being, although it moves away from coming into being and existence. When something is proceeding towards existence and being, then we say that it is coming into existence and being, not that it is perishing, although it is moving away *from* perishing, as it is moving *towards* being.

224b11 Forms, affections and the place towards which moving subjects move, are immovable.

Yaḥyâ:[6]
He uses the word 'motion' instead of 'change', and he takes knowledge as an example of a form, because in the book *On the Soul*[7] it is said that the ways which lead to knowledge are related to change and to generation in particular.[8] The word 'form' also applies in the case of growth, because growth ends at a certain form; the word 'affections' applies to the case of alteration.

Yaḥyâ:
[224b13-26] He brings up a problem, namely that forms must be motions. For forms, such as whiteness and others, are affections (*aḥdât*), affections are motions, so forms are motions. The solution is that it is not forms which are affections, but the process towards a form. This process is also a motion, not a form. Furthermore, if affections are motions, and were also to be forms, then they would have homonymous names.[9] Then one would call a form, such as whiteness or blackness, an affection, where 'form' is said of that which is completed, and one would also call the process towards the form an affection. If it is a homonymous word then the premiss which says that the affections are motions is not acceptable.[10] We say, however, that considering the forms to be affections is not admissible. A similar homonymy occurs in what Socrates says in *Phaedo*,[11] namely that the contrary things do not arise from each other; he also says that the contrary things do arise from each other. This is not a contradiction, because by 'the contraries which do not arise from each other' he means the forms, such as blackness and whiteness. For these do not come into being by arising from each other, except in the sense of coming into being one after another. The statement that the contraries come into being by arising from each other concerns that which is composed of matter and form, such as what is white or black, for such things arise from each other.

Then Aristotle states that the motion which exists on account of a part and by accident may occur in all kinds of motion. This statement also explains his remark at the beginning [of this chapter] that 'it may be different [for each kind of change]'.[12] He presents as an example the alteration which is accidental for the colour white when it changes into an object of thought. For when white changes into black and someone is thinking of the fact that it has become black, then this alteration to an object of thought does not occur in virtue of itself, but it is accidental that it has become an object of thought. Something similar holds for an alteration from black to white on account of a part: when we say that the black changes to a colour, then its change to a colour occurs on account of its change to white, and white is a part of colour. In the same way an alteration in virtue of itself is exemplified by the statement: 'White and whiteness change into black.' Similarly, a local motion on account of a part is referred to in the statement: 'I have moved to Iraq', when I have moved to Basra, because Basra is part of Iraq.[13] Similarly, a local motion by accident is meant in the statement: 'The white goes to Basra', because the one who went did not go insofar as he was white, but he was only accidentally white. Similarly, a local motion in virtue of itself is: 'Zayd went home', because this belongs to him primarily, not on account of something else, namely a subject or a part. Motion in the sense of growth may be classified in the same way.

He makes a distinction between the expressions 'in virtue of itself' and 'accidentally' and between 'primarily' and 'on account of something else,' such as motion by a part. Motion on account of something else may be on account of a subject, or on account of a part.

Yaḥyâ and Abû 'Alî:

500,14 [224b26-35] Having finished the discussion of the classification of motion he tells us what he is going to discuss, saying that he will not talk about accidental change, because this occurs in all species of each of the ten categories, and in this respect it is not determined, and
501,1 science does not investigate what is not determined. Essential change occurs only in those categories in which there are contradictories and contraries; therefore it is determined and so may be investigated by science.

In explanation of this passage, we may say about the generation of a substance that a man is generated from the white, because white is an accidental [attribute of] the sperm. When we say of a magnitude that two cubits arise from the white, this occurs accidentally, because it is accidental to the thing which becomes two cubits that it is white. We may say about a quality that white comes into being from hot, and we may put forward similar statements for the other categories. Thus, as accidental change is not restricted to change between contradictories and contraries, there are many kinds of it and it occurs everywhere. But since essential motion is confined to these two [kinds], it is not ubiquitous, but restricted.

An example of change between contradictories is when a thing changes from something not existing to something existing and from not-white to white. Change other than coming to be,[14] however, [even] if it is between contradictories, is change from one contrary to another, for example the change from not-white to white, because not-white is either black or one of the intermediate colours. 'Something not existing' and 'something existing' are not contraries. Change between contraries is like the change from white to black; change between intermediate states is like change between contraries, because in such states there is something of the contaries, e.g. in grey there is something of white and something of black, although neither of them is really in it. Likewise, if one compares the intermediate state to each of the extremes, it is a contrary to [both of] them. For when grey is compared to white it is black, and when it is compared to black it is white. Therefore a change from grey to white is a change between contraries.

224b27 Because it occurs in everything, always, and in any respect.[15]

Yaḥyâ:[16]

Accidental change occurs in all things, because it occurs in every 501,22
species of the categories, for it is in heaviness and lightness, in heat and in coldness and in other kinds of quality. His statement 'It occurs always' stands for 'We may find this in each of the categories.'

{224b35-225b5: Motion or change[k] (*kinêsis*) is a species of change (*metabolê*) which proceeds from one intermediate or contrary to another and involves an enduring subject of change.}

Yaḥyâ:

Having discussed the preliminaries needed to [explain] the difference 503,10
between motion and change, he now begins to explain the difference by saying that generation is a change, not a motion. Before that he points out what change is and how many kinds of change we may distinguish. Then he passes on to his goal: the difference between change and motion.

Change is the process from one thing to another. The Greek word [*metabolê*] gives an indication of this, for it implies that something comes after something else and goes towards something. If change consists in one thing becoming another, then one may distinguish four kinds: (1) Change from a subject to a subject – where 'subject' means something existing and what is positively expressed, and

non-subject something not existing and what is negatively expressed. (2) Change from a subject to a non-subject. An example of the first kind would be change from white to black and an example of the
504,1 second kind would be change from a human form to another one.[17] (3) Change from from a non-subject to a subject, such as change from sperm, which does not yet have the form of a man, to a man. (4) Change from a non-subject to a non-subject; this [kind] does not exist, because change is always between contrary or contradictory oppositions, and a non-subject can by no means be opposed to another non-subject.

Change from subject to subject occurs when the accidents change, because the subject remains in the same state during the change of its accidents; an example is the change from white to black.

Change from non-subject to subject is generation; the subject does not remain in the same state, because the change occurs in the substance, not in an accident; an example is the change from not-man to man.

Change from subject to non-subject is corruption; this is also a change in the substance itself; an example is the change from man to not-man.

225a14 Unqualified change is unqualified generation; a particular change is a particular generation of something.

Yaḥyâ:[18]

Unqualified change is unqualified generation; unqualified generation is generation of a substance, because it is substance which is generated without qualification, as the change occurs in substance. A particular change is a particular generation; a particular generation is a generation of an accident, because it is an accident, not a substance which is generated: when something changes from black to white the substantial form remains unaltered, whereas it takes off the form of blackness and puts on the form of whiteness. Similarly, unqualified corruption is corruption of a substance, and a particular corruption is a corruption of accidents.

225a19 A particular corruption is a change to the opposite negation.

Yaḥyâ:[19]

505,2 What is corrupted from white is corrupted towards its opposite negation, i.e. not-white; that towards which it is corrupted is necessarily black or one of the intermediate colours.

Yaḥyâ:[20]

Aristotle explains that generation is not a motion. He proceeds to explain this by starting from a classification of what does not exist. What does not exist may be classified according to the classification of what exists, because it is its opposite. One may say that something exists in different ways: (1) Something may exist as a statement, and describing a statement as existing means asserting that it is true; a true statement can only be composite, because separate words cannot be true or false. (2) Something may exist in the absolute and proper sense; this is the existence of a substance, i.e. the existence of the form by which the composite [of matter and form] is what it is. This is existence in the absolute sense, because it exists on account of the fact that it does <not>[21] cease to exist when something else, such as whiteness or blackness, ceases to exist; whereas something else ceases to be when it ceases to be, since when the form of a body no longer exists, none of its accidents continue to exist either. (3) Something may exist accidentally, which is a particular existence, e.g. a body's being white or black. 506,2

What does not exist may also be classified in these ways. (1) What does not exist as a statement: these are the false statements. They cannot be in motion: the belief that something is the case may be wrong at one time and correct at another time, such as our belief that Zayd is at home. This is not because the belief is subject to motion, but because that which is believed is subject to motion. (2) What does not exist in an absolute sense: this is what does not exist as a substance, for instance something which does not exist as a man. (3) What does not exist in a particular sense, such as the non-existence of accidents: this occurs when a body does not have a certain accident. We call this not-being in a particular sense, not not-being in the absolute sense.

He explains that generation is not motion in two ways. First, that what moves must necessarily exist, and prime matter, which is changed and receives the form [when a generation occurs], does not exist in the absolute sense, because it is not something which exists completely and actually. Thus prime matter cannot be in motion. Secondly, all things that move are in a place, and since prime matter is not in a place, it cannot move. Thus he refutes the claims that generation is motion and that a statement could be subject to motion.

225a21 For example, what is said [not to exist] in the way of conjunction or separation[22] cannot be in motion. 507,1

225a22 What is said [not to exist] in the way of being potential,

> which is the opposite of what exists absolutely, cannot be in motion either.

Yaḥyâ:
What exists potentially is opposite to what exists actually. 'Potential'[23] is said of the prime matter and also of the form, because we say that the not-white is potentially white. As the potential may be divided into these two classes and Aristotle means prime matter, he adds to his statement the words 'which is the opposite to what exists absolutely and actually'. Because prime matter is not something which exists perfectly and actually, it cannot be in motion. The not-white can be in motion accidentally, e.g. when the not-white is a man, because that man exists actually, and such things can be in motion.

> **225a25** What does not exist absolutely is not a certain 'something', therefore it can in no sense be in motion.

The word 'absolutely' is used to refer to what is not a substance and not one of the things that exist actually.

> **225a26** If that is the case, then generation cannot be motion, because it is the non-existent which is subject to generation.

Yaḥyâ:
What will be existing is understood to be not yet existing, and what does not exist cannot move.

Yaḥyâ:[24]
508,9 Aristotle has said in the first book of this work[25] that things may be divided into those which exist by themselves and those which exist accidentally. What exists accidentally is the privation, because in itself it is non-existent; that is the way in which it is understood. As it is an accident for a subject, however, and the subject exists, it is said that the privation[26] exists accidentally, and also that it changes accidentally, in virtue of the changing of its subject.

As for the things which exist by themselves, one may classify them into those which exist in that way only, such as the substantial forms (for they exist by themselves and cannot be the subject of contraries, so that they could be accidentally non-existent in virtue of their being a subject for privation) and those which exist by themselves and are accidentally non-existent, such as prime matter which exists by itself. As privation is connected with it, one says that prime matter is accidentally non-existent, because the privation does not belong to it essentially; for if it did, it could not receive a form. If prime matter exists by itself, and it is that which is subject to generation, then an opponent could object to Aristotle's statement that prime matter is non-existent, and therefore cannot be in motion. The answer is that although prime matter exists by itself, privation is connected with it, and this privation is non-existent by itself. Since the prime matter receives the form in virtue of its privation, generation is not a motion. 509,1

> **225a28** It is nevertheless correct to say that it is the non-existent which is subject to generation in the unqualified sense.

He refers here to prime matter. The expression 'in the unqualified sense' means: in the proper sense and by itself.

> **225a29** The same is true for being at rest.

He means that what he has mentioned before, namely that what does not exist cannot be in motion, holds in a similar way for being at rest, because what does not exist cannot be at rest, as resting is the termination of motion, and what does not move cannot have a termination of motion.

> **225a30** Similar absurdities arise ...

I think that he means that if prime matter were to be at rest, then absurd conclusions would follow, similar to those resulting from the statement that prime matter were to be in motion, namely that what does not exist would be at rest or in motion, so that when its motion stopped it would be at rest.

> **225a32** Corruption is not a motion either.

If corruption were to be a motion then its contrary would be either a motion or a rest; the contrary of corruption is generation, and generation is neither a motion nor a rest.

Yaḥyâ:
Having shown that certain kinds of change are motion, but others not, Aristotle now wants to give a definition of motion. In the third book of this work there is a definition: 'Motion is the actuality of what exists potentially, as far as it exists potentially.'[27] This definition includes not only generation[28] but also all the other kinds of change. Because this definition includes all kinds of change he will now give a special definition of motion, which does not apply to the other kinds of change. This definition is that motion is a change from one subject to another. This is equivalent to our statement that it is a change 510,7

from one contrary [to another] and a change from one actual form to another.

There are three kinds of change: generation and corruption, change from one contrary to another, and from one magnitude to another. The word 'motion' should not be used for generation and corruption, because these processes imply contradictory states; it is rather used for changes between one subject and another – and by 'subject' I refer to what is positively expressed. This may be either a contrary, such as [occurs in] the change from black to white, or an intermediate state, such as grey.

Because he regards the intermediate states as privations someone might say that if a change from or to an intermediate state is a motion and moreover a motion from one privation to another, then why is a change from [a state of] corruption [i.e. not-being] to being not a
511,1 motion, since that is a change from privation to form? He [Aristotle] says that grey and other intermediate states are privations in the sense that they are not the extremes, as grey is neither white nor black. However, compared to the extremes they are contraries, because grey is black if it is compared to white, and compared to black it is white.

That these states are like privations becomes clear from the fact that they have special names, and may be positively expressed. For instance, grey has a special name and is positively expressed. 'Naked', as a privation, behaves in a similar way; we refer to it by a positive expression, as our word 'naked' is not negative, and that word is also a name. The contraries themselves are privations, because it is said that white is not-black, and black is not-white. One refers to them, however, with positive expressions, i.e. 'white' and 'black'. The situation is similar for the intermediate states.

{225b5-16 Because of the nature of motion (change[k]), it can only occur in relation to the categories of quality, quantity and place.}

Yaḥyâ and Abû 'Alî:
512,7 Having shown which kind of change motion is, Aristotle wants to show in which categories motion occurs. He enumerates seven of the ten categories; he omits the categories of 'having', 'being in a position' and 'when'.[29] Then he explains in which category motion occurs; it appears that there are three such categories.

He does not mention the categories of 'having' and 'being in a position', although it may be thought that change occurs in them, because a subject may at first not be 'having' something and then be 'having' it: Zayd may be not reclining [and then reclining]. He does not mention these categories because both fall under the category of 'where'. For the category 'having', which has the sense of possession,

admits of proceeding from one place to another and this motion falls under local motion. What is 'in a position', for instance the reclining 513,1
person, admits of the motion towards the place at which he is reclining from the place at which he was standing.

No change is possible in the category of 'when', because according to him time is one and continuous, and in time there is no opposition, nor does it stop. Because this is evident he has not mentioned it.

[Aristotle 5.2]

In the category of 'substance' there is no motion, because a substance has no contrary.

Turning[30] to the category of 'relation', change[k] does not occur for either of the related things, although the relation between them may become different.[31] For example, what is on my right may change its place and come to be on my left, so that I am then on its right without myself having been subject to a motion or change. Furthermore, that which comes to be on my right acquires this relation accidentally. Its motion by itself is its motion to that place, and it is accidental to that place that it is on my right.

As for 'action' and 'being acted upon' ('passion'), if there were to be change[k] in these categories, then it would occur from one action to another and from one passion to another, in the same way as change[k] in the category of quality occurs from one quality to another, like the change[k] from coldness to heat. Action and passion are the same in subject, but different in definition, because what arises from the agent is the same effect as what is received by the patient. Therefore they are the same in subject, but in relation to the agent it is called action, and in relation to the patient it is called passion, so they differ in definition in accordance with these two relative terms (agent and patient).[32] If action and passion are the same in subject, and action and passion are changes[k] and change[k] is not subject to change[k], and in general change is not subject to change, then there cannot be change[k] in the category of action and passion. Therefore the mover is not subject to change[k], i.e. a change[k] by which it would start the action of change[k], nor is the moved subject to change[k], i.e. a change[k] by which it would acquire or part with the form of being moved.

{225b16-35 If there were to be motion (change[k]) in relation to the categories of action and passion, then it would have to be possible that there be motion of motion. But there cannot be motion of motion *per se*.}

Yaḥyâ:
His intention here is to show that motion cannot be subject to motion. 515,11

He says that motion might be subject to motion in two ways: either by way of being subject to another motion, as a man may be subject to getting healthy or getting ill, or by way of not ending at a form and not stopping at a state of rest and standing still, but instead stopping and ending at another motion, as if getting healthy ended at getting white.

It is not possible for a motion to be subject to another motion at all, because motion cannot be a subject for anything, as every subject has a form, and should be able to exist by itself. The concept of motion implies that motion is always in something and that motion can in no way exist by itself. It also implies that motion ends at a certain form, and motion itself is not a form.

As for the statement that a motion does not arrive at another
516,1 motion, if one says that it arrives at another motion accidentally, then this is possible. But we have said that in our discussion on motions we do not consider accidental motions. Something may accidentally arrive at something else, as for instance [when] a man is born out of sperm [one may say that] something white is born out of sperm. The sperm does not arrive by itself at the white, but by itself it arrives at the man, and white is an accident of the man. Similarly motion may accidentally arrive at another motion, as for instance learning by itself ends at knowledge of something and may accidentally end at forgetting something else, as the person who is learning something, or devoting his atttention to it, may accidentally forget something else.

It is not possible that a motion arrives by itself at another one, and in general this is not possible for any change, regardless of whether it is generation, corruption or motion. For if a generation were to arrive by itself at another change, then it would have to arrive at a change which is opposite to it, because a motion arrives by itself at the opposite; otherwise it would not arrive there rather than somewhere else. The opposite of generation is the process towards corruption, so it follows that the more something is generated, the more it is corrupted, insofar as it is generated, for, if something proceeds to something, the more that which proceeds to something comes to be, the more that to which it proceeds comes to be. It would follow that the more existent something becomes, the less existent it becomes.

The discussion of motion is similar. If the process of becoming healthy were to arrive by itself at some other motion, then this motion would have to be the opposite of becoming healthy, namely becoming ill. Therefore, the more healthy a person becomes, the sicker he gets, and this is absurd. Therefore there is no change of change in this way.

Furthermore, what moves, moves towards a goal at which it stops, because if it did not move towards a goal at which it stopped, but continued to move without a goal, then its motion would be in vain.

Someone who is ill must necessarily move to a state either of corruption or of health. If he is moving towards corruption, and the change changes towards the opposite, in accordance with what has been posited here, then the one who is being corrupted is becoming healthy, and if he is proceeding to becoming healthy,[33] then the sicker he is, the more healthy he becomes. Therefore a change cannot be subject to a change.

225b21 The other way is that some other subject might change.[34] 517,1

Yaḥyâ:
He means to say that motion cannot be a thing[35] and a subject [of motion], but the subject might be something else, which could change from one change to another kind of change; the word 'form' should be understood as 'change'.[36] Thus he means to say that motion is not a subject [for change], but the subject for the change is something else, and that change might arrive at another change. That is possible only accidentally.

225b24 Because motion itself is the transition from one thing to another.

Yaḥyâ:
He says that it is not possible that a motion by itself arrives at another motion, because the concept of motion implies that it is a transition from one thing to another … .[37]

If a motion were to arrive by itself at another one, then it would not be true that motion is a transition from one thing to another.

His statement 'The same is true for generation and perishing' [225b25] means that what holds for motion, holds for generation and perishing too, i.e. that they are transitions from one thing to another.

225b26 Then what changes would change from health to illness and at the same time be changing from this very change to another.

Yaḥyâ:
He says that if a change were to arrive by itself at another one, then it is clear that it should arrive at a change opposite to it; therefore becoming healthy would arrive by itself at becoming ill. If something arrives at a certain thing (goal) then with each part of the motion a part of the goal must be realised, so that if the motion is finished, the goal is complete, just as we say of generation that with each part of it a part of the form has come to exist, so that when the

generation is finished and has come to an end, the form is completed. It follows that becoming ill begins simultaneously with becoming 518,1 healthy, so that becoming healthy exists and does not exist at the same time, for if becoming ill exists together with it, then becoming healthy cannot exist. It follows that when the process of becoming healthy is finished, the process of becoming ill is also finished, and that health exists because the process towards health is finished and illness exists because the process of getting sick is finished, so that a man would be ill and healthy at the same time. This absurd conclusion follows from the supposition that it is possible for the process of becoming healthy to arrive at health and also at becoming ill.[38]

> **225b28** It is clear then, that when he has become ill, he will also have changed to some other change, whatever that change may be, for he could also be at rest.

Yaḥyâ:
This might be another proof, namely that what moves and changes always ends in rest. Suppose getting healthy ends at health and at the same time at some other change, whatever it may be. It follows then that that at which the motion ends is a motion on the one hand, namely proceeding toward illness, and a rest on the other hand, since the process of becoming healthy has come to an end. If this were so, then rest would be the same as motion.

> **225b29** Furthermore, this change can never be any arbitrary change, but it must be a change from something to something; so it must be the change towards the opposite change, which is becoming healthy.

Yaḥyâ:
When he said before: ' … . he will also have changed to some other change, whatever that change may be … ', he forced them to a worse absurdity than the first one. The first absurdity does not imply that the change at which the change ends is an opposite change. The second absurdity is that becoming healthy and becoming ill exist together, because if something proceeds toward something else, then necessarily a part of the goal becomes realised with each part of that process; this has been explained before.

Yaḥyâ:
519,1 I have considered the passage where in one part he says that the change ends at any arbitrary change and in another part he says that the change does *not* arrive at any arbitrary change, as two proofs. I think, however, that it is one proof, which runs as follows: If change

has a change, then either it arrives at any arbitrary change, or it does not arrive at any arbitrary change, but at an opposite change. The first alternative is impossible; the second one is possible. Because what moves may come to rest – as we are concerned here with rectilinear motions – becoming healthy must arrive at health, for health is the coming to rest of becoming healthy, and because getting sick occurs at the same time as getting well, as has been explained, it follows that he has become ill when it stops, i.e. when becoming ill stops. Thus getting sick stops when getting well stops, and at that moment health is realised; it would follow that when health is realised, illness is also realised.

> **225b31** It can be only accidentally, however, [that there is change of change]; an example is the change from remembering a thought to forgetting it; this is a change belonging to the subject who has this thought: at one time he changes to knowledge, at another to ignorance.[39]

Yaḥyâ:[40]
He says that it is not possible that there is motion of motion, nor can one thing simultaneously be subject to two contrary motions, except accidentally and from two points of view. An example is that the very same man, while he is remembering one specific thing, at one specific time, could be forgetting something else. He does not remember that thing on account of that by which he is forgetting, nor does the remembering cause him to forget, but it is accidental for him that he forgets what he forgets while he is remembering what he remembers.

This is similar to walking, which results in becoming healthy or hot. This does not occur by itself, but because when someone moves, this is connected to a latent heat which is aroused[41] by the motion; in the same way the motion of the limbs makes them hot while it arouses the heat which is inside the body.

Yaḥyâ:[42]
He says that this opinion implies an infinite regress, because if there were motion of motion it would follow that a present motion must be preceded by another motion, and similarly this preceding motion must have another motion, and this continues without there being a first one. If this series of motions were infinite and did not have a beginning, then a present motion would not exist, because it can exist only because the first one has existed, and if an infinite series does not have a first term, the next and following ones do not exist either. 520,5

Yaḥyâ:
Then Aristotle says: ' … or a generation of generation.' He turns to

generation after having mentioned its genus, namely change, because the species in each genus are more clear than the genus. Similarly the explanation connected to the species is more clear. Therefore he mentions generation after change.

Yaḥyâ:

522,14 [226a1-6] Aristotle starts his argument by assuming he is discussing absolute generation. By 'absolute generation' he means here something different from what he usually means, i.e. the generation of a substance; here he means simple generation,[43] whether it be generation which has still to occur, or a past generation, or generation of an accident.

We say that if absolute generation, whatever it may be, were being generated and were to have a generation, then it would follow that there is no generation at all. This would imply that nothing would
523,1 ever have come into existence, because if the generation$_1$ of a thing is being generated$_2$ [and this generation$_2$ is] at a certain stage, then at that stage the generation$_1$ is not a complete generation. It is a complete generation when it exists and its generation$_2$ is finished; it has then become a generation. At a certain stage of its generation$_2$ it is not yet a generation. The generation$_2$ of the generation$_1$ may also be in a certain stage of generation$_3$. If its generation$_3$ is not a completed generation,[44] then the generation$_2$ does not exist, because
5 the first generation is the cause of the other and if the cause exists, what is caused by it also exists. If what is caused does not exist we know that its efficient cause does not exist either. If the cause does not exist, what is caused does not exist either. Thus, [the existence of] this generation is dependent on the fact that it has a generation, and its generation is [in turn] dependent on its having a generation which is its cause, and this continues *ad infinitum*, so that every generation is preceded by a generation. An infinite series has no first term, so the series of causes of this generation does not have a first
10 term, and as the first term does not exist, the following terms do not exist either. On the strength of these assumptions simple generation would not exist, and there would be no generation of things.

Yaḥyâ:

As an example of this argument one may refute the claim that Socrates has an infinite number of ancestors, because the infinite has no first term, and when the first term does not exist the following ones cannot exist either.[45]

Yaḥyâ and Abû 'Alî:

524,11 [226a6-16] Aristotle presents another proof, which runs as follows: Every rectilinear motion goes from one contrary to another, because what moves does not move from one arbitrary thing to another arbitrary thing, but from one contrary to another. This statement also includes the other kinds of change, such as generation and others. Something may change from one contrary to another, while the subject does not stay in the same state. For fire does not change from hot to cold while the subject of the heat stays in the same state; this does not, however, invalidate the statement that the contraries between which the change occurs belong to the same genus. Furthermore, rectilinear motion must have a goal; if not, then the action of nature would be in vain,[46] if it failed to bring motion to an intended goal. If motion has a goal, then its goal is rest, for if something moves towards a goal, this implies that it proceeds to the goal in order to 525,1 reach it. If it has reached the goal, its motion stops, and the stopping of the motion means coming to rest.

Thus it is established that change goes from one contrary to another. If a generation were subject to generation, then it would also be subject to perishing. Then the generation would perish either before it had started, or after the generation, when the process was finished, or during the process of generation. It is impossible for a generation to perish when it has not yet come into existence, because then what cannot be perishing would perish. It is also impossible for it to perish when its generation is finished, because when the generation is finished the situation has returned to its orginal state, before its generation. Thus it must perish during the process of generation, but then the generation would perish while the process of generation is going on.

This is not to be taken in the sense that we may say that part of a generation is completed and part of it still has to be completed, and what has been completed has perished and what has still to be completed has not perished, because we assume that the discussion is about what is completed, when we say: Do you think that it has perished before the generation has started, or after its completion, or during the generation? The discussion of each part indicated by the questioner runs in a similar way.

We say that the generation of a thing does not arise or disappear in a period of time, and that it is not generated, because generation is not a subject.

If generation is not subject to perishing, then it is not subject to generation either, because what is generated towards something changes towards its contrary or contradictory.

Aristotle says: 'It is necessary that it is at rest, and at the same time it is a motion.'[47]

He draws this conclusion because if a motion necessarily has a goal, this goal is a state of rest. Then the supposition that a motion has a motion implies that the motion is rest, for the motion [of the motion] has arrived at its goal and also at a motion. Insofar as it has arrived at a goal it must have arrived at a state of rest, and insofar as it ends at a motion the goal must be a motion.

526,1 **226a6** Furthermore, the same subject may have the contrary motion, and also the rest, and [the same subject] may be generated and perish.

Yaḥyâ:[48]
In this passage he gives a proof of the proposition that a change does not have a change, as follows: If a subject changes to something, then the same subject is able to change to the opposite. Thus if generation had a generation, then it would also have a perishing. Then it would perish either before the generation, or after it, or during the process of generation. It is impossible for it to perish before the generation, or after it, or during it. It is impossible that it would perish after the generation, because it has no stationary point at which it ends, so that one could say that it started to perish from there; it is always in motion without getting to a stationary point. It is also impossible that it perishes while the generation is going on. Thus there is no generation of generation.

226a7 ... and also the rest.

Yaḥyâ:[49]
He did not add this because it is necessary for the proof, but because it is something which necessarily follows, because what has a certain motion as well as the contrary motion must be at rest at a certain time. What is at rest at a certain time in opposite resting states is in opposite places. Because the places are opposite, e.g. above and below, the rests are opposite and the motions are opposite.

226a12 Furthermore, what is generated and changed must have an underlying substrate.

Yaḥyâ:[50]
This is another proof, namely that every change occurs in something underlying it, and everything which changes is *in*[51] something which is different from that to which it changes. Matter, for instance, is different from what it becomes, e.g. it is different from the horse [which is generated from the matter], and the wood is different from what it becomes, namely a chair. What is generated and changed, when it is generated and changed, becomes another substrate. Therefore, if a generation were to be generated, then before its generation there must exist some underlying substrate, and this substrate will

become some other substrate when the generation occurs. That is not possible for generation, because it is not something stable of which one might say that it is something different from what it becomes, and that it becomes something different.

226a13 Furthermore, what would be the goal towards which the motion [of motion] occurs?

Yaḥyâ:

Just as everything which changes must have a substrate, its change must also arrive at something which is different from change itself. 527,1
Then if a change has a change then the first change must have a goal which is not change, and similarly for the second change. If it is the preceding change which exists, then the second one cannot exist, because the goal of the first change is different from change. If it is the second change which exists, then the first one cannot have existed, because the existence of the second one makes the existence of a preceding change superfluous, and if it was preceded by a change, then that one would arrive at something which is not change.

226a14 And also there is no way to conceive how this could be possible, because the generation of learning is not learning.[52]

Yaḥyâ:[53]

He says that if every generation had a generation, then generation would not have a certain goal; therefore learning would also have a generation, because learning is a kind of generation. If learning had a generation then this generation would be a learning, just as the change of a motion is a motion. It is impossible that learning has a learning, because everything which is the result of learning is something which one knows and has learned; whereas learning is not something which one knows and has learned, but rather it is a course and a way towards what one has learned.

Yaḥyâ and Abû 'Alî:

[226a16-23] This is another proof of the statement that change[k] does not have a change[k]. For change[k] occurs in three categories: quantity, quality and the 'where'. The change[k] of a change[k] must fall under one of these categories and if change[k] changes[k] it must change[k] towards one of these. Let us suppose that the change[k] which is subject to change[k] is a change[k] in the category of the 'where', and let it be walking. If walking is subject to change[k], it must change[k] towards one of these changes[k], i.e. a change[k] in the 'where', in quantity, or in quality. It is absurd that local motion and walking would move to a quality – for then walking could become cold or hot. It is also absurd 528,5 that it would move in quantity, for then the walking could grow or diminish. Furthermore, it is absurd that local motion would be subject to local motion, for if we suppose that the local motion is walking, then walking would walk, and the walker would be walking on account of the fact that [his walking] walked, because walking would be a subject for walking.[54]

By the expression 'the underlying nature' [226a17: *tên hupokeimenon phusin*] he means the change[k] which is supposed to be subject to change[k]; by the expression 'that towards which the change[k] proceeds' [226a18: *eis ha kinountai*) he means one of these three changes[k]. So he means to say that the change[k] which is subject to change[k] must move towards something in one of these three categories. He illustrates his intention by saying: 'local motion would be subject to alteration or local motion' [226a18].

It is possible that a change[k] is subject to a change[k] in an accidental way. For it is possible for a man to be learning while he is being cured. Then it is accidental for him that he is learning while he is being cured. But accidental change has been left out of Aristotle's account.

Yaḥyâ:

529,6 [226a23-6] He summarises the result of his discussion above, namely that there is no change[k] in the category of substance, nor in those of action, passion and relation. Then he says that change[k] does exist in the categories of quantity, quality and the 'where'. He explains this by what he says after that, namely that change[k] occurs between contraries, and contraries exist in these categories only; therefore change[k] occurs in these categories only.

> [**226a26-7** Let the change[k] in respect of quality be 'alteration', a common term which combines both contraries.]

Yaḥyâ and Abû 'Alî:

This means that the word 'alteration' is a term which is used to include all changes occurring in the category of quality.

Yaḥyâ:

The word 'combines' [226a27: *epezeuktai – qurina*] is used as an equivalent of 'posits' (*wuḍi'a*).

Yaḥyâ and Abû 'Alî:

530,6 [226a27-9] Having said that the word 'alteration' includes all changes occurring in the category of quality, while the substantial differentiae also fall under the concept of quality – Porphyry has said that the substantial differentiae indicate what kind of thing something is, because they designate distinctive qualities in a substance – he means to explain that the word 'alteration' does not refer to the substantial differentiae, but to the accidental differentiae in virtue 10 of which a thing is capable or incapable of being acted on, such as heat or sweetness, as the substantial differentiae are not subject to change[k].

Yaḥyâ and Abû 'Alî:
[226a29-32] This means that change[k] in the category of quantity has 531,2 no name inclusive of all kinds, as is the case for the kinds of change[k] in the category of quality. Each kind of [change[k] in] quantity has a special name, such as growth or diminution. Each one of these names designates each of these kinds.

Yaḥyâ and Abû 'Alî:
[226a32-b10] He says that the kinds of change[k] in the category of 13 place have a name common to all of them, namely local motion. He says that the term 'local motion', however, refers in particular to the things which change place and do not have the power to stop or are not moving by themselves. He refers here to the forced motion of a stone and such kinds of motion, because the stone cannot stop before the end of its motion and its motion is not by itself.

Yaḥyâ and Abû 'Alî:[55]
As he said that the name alteration includes all kinds of change in 532,6 quality, and since change in quality occurs from one contrary to another, alteration necessarily goes from one contrary to another. Aristotle is aware of a problem here which has the following form: We find that an alteration may occur from a greater to a lesser degree and from a lesser to a greater degree.[56] An example of an alteration 10 from greater to lesser is that an intensely black colour becomes faintly black, and alteration from lesser to greater is that a faintly black colour becomes intensely black. The solution of this problem is as follows: Such an alteration is also between contraries, because it is an alteration from and to an intermediate state, and the intermediate states have in them something of both contraries. Thus, if the alteration is from less to more, such as the alteration from a faint black to an intense black, it is an alteration from a contrary, in the sense that 15 it is an alteration from what is contrary in a lesser degree to what is contrary in a greater degree. And if the alteration is from more to less, then it is an alteration from what is contrary in a greater degree to what is contrary in a lesser degree.

Yaḥyâ and Abû 'Alî:[57]
He says that if a change which occurs from an intermediate state to 533,8 an extreme, it goes from the part contrary to the extreme which is

present in the intermediate state; from this contrary the change occurs. An example is the change from weakly white to intensely white; this is a change from the black which exists in the weakly white. For a weakly white colour only exists in what has some parts of blackness in it. Change from an extreme, e.g. from an intensely white, to an intermediate state is also a change from a contrary, namely white, to a contrary, because in the intermediate state, parts of black are mixed in. Thus change and alteration in quality are always from a contrary to a contrary.

Yaḥyâ and Abû ʻAlî:

534,11 [226b10-17] Having explained the difference between motion and change, that not every change is a motion, and that motion occurs in three categories, he now starts to enumerate the things which are called motionless.[58] Since being at rest is a kind of opposite and contrary of being in motion, and since an opposite exists when its opposite no longer does, it is necessary that rest exists in what has no motion, but is capable of being in motion; for if something is not capable of being in motion, it cannot be a subject for it, and so it cannot be a subject for rest either, since the subject of two contraries must be one and the same thing. Therefore enumerating the things which are motionless is useful in order to know what things are at rest.

Something may be [called] motionless when it is not subject to motion and is incapable of being in motion, for example the things
535,1 which are not bodies. One may also call something motionless when it is moving slowly, as we say that the fixed stars are motionless. Furthermore, something may be said to be motionless, when it can only be put in motion with difficulty and hard work, although when it has started to move the motion becomes quicker. An example is a young raven, whose motion is slow, i.e. it is slow in starting its motion, unlike a young pigeon. When the raven's motion becomes quicker, it is not inferior in velocity compared to the motion of the [adult] raven. Finally, something may be called motionless when it is not in motion, but admits of motion, so as to be motionless at the time and at the place at which it admits of motion, and in the manner in which it admits of motion. For example, when Socrates is sitting we say he is motionless, for he admits of motion at that time and at that place, but he has stopped functioning as he would during his motion. When he is walking we do not say that he is motionless, in the sense that he is not moving in the air, because he does not admit of motion at all in that place (sc. the air); we do not say he is motionless, in the sense that he is not flying, because he does not admit of motion in that way. Nor do we say of a child who has just been born that he is motionless in this sense of 'motionless', because at that time he is not able to move at all.

This [last] class of motionless things, with these conditions, is said to be at rest, because if rest is the opposite of motion, the subject of both must be the same. What admits of motion fulfils all these 20
conditions, so only such things admit of rest.

226b11 ... and what can only be moved in a long time with great effort.

Yaḥyâ:
He says that a thing is slow to move when its motion starts only with great effort or is slow to be moved.

226b15 Rest is contrary to motion, so it is the privation [of motion] in that which admits of motion. 25

Yaḥyâ:
He says that rest is the privation of motion, because the inferior of 536,1
two contraries is called the privation; rest and motion are contraries, and the inferior of them is rest. Therefore he calls it a privation, whereas he also calls a privation a contrary. He does this at many places from the beginning of this study.

226b16 From what has been said it has become clear what 11
motion is.

Yûḥannâ:[59]
Namely change from one subject to another.

226b16 ... and what rest is.

Yaḥyâ:
It is the privation of motion in what admits of being in motion, at the 15
time and at the place at which it admits of being in motion, and in the way in which it admits of being in motion.

226b16 ... and how many kinds of change there are.

Yaḥyâ:
Four kinds: change in substance, quantity, quality and the 'where'.

226b17 ... and what kinds of change[k] there are.

Yaḥyâ:
[There is change[k]] in quantity, quality and the 'where'. It is also clear

that rest also exists in these categories, because changes[k] exist in them.

[Aristotle 5.3]

Yaḥyâ and Abû 'Alî:

537,9 [226b18-23] He intends to discuss the meaning of the expressions 'touching', 'between', 'in succession', 'contiguous' and 'continuous', because he needs to know the meanings of these words for what he wants to talk about afterwards, for he will discuss what a continuous motion is and when motions are one.

Some of these terms can be true only of bodies, and others can be true as well of what is not a body; for instance 'between' and 'in succession', because 'between' may be said of the colour grey, as it is between white and black, and the number two is said to succeed the number one.

Yaḥyâ and Abû 'Alî:

538,4 Things may be 'together' in time and 'together' in place. He intends here to discuss the 'together' in place, because he wants to talk about motion. Therefore he specifies this by saying 'together in place'. Things which are together in place are those which are in one and the same proper place. He says 'proper place', because place may be a proper place, i.e. the boundary of the surrounding body which is equal to the surrounded body, or it may be a non-proper place, such as a room, for one who is in a corner of a room is in the room. One may also say that he is in the house, and in the town. We do not say that Zayd and 'Amr are together in one and the same place, if they are both [merely] in the same town. We say that they are in the same place when one place surrounds them, in the way that water is in a mug and all the parts of the water are together in the same place (the mug). This only occurs when the boundaries of the parts are not separated. Thus we do not say that the thin, flat loaf of bread which is spread beneath the food (*aṣ-ṣabîr*) is in one place, because each grain is separate from the others.

What is together in time is that which is in the same time, such as one day or one week; we do not mean a proper time, because all time has a certain 'width' (extension). Things are 'apart' when they are in different places, like Socrates and Plato.

[**226b23** Things are said to be touching when their extremes are together.]

Yaḥyâ:[60]

539,2 If by 'extremes' he means parts, then they cannot be together, because they cannot be in the same place at all. If he means the boundaries, then they cannot be in a place, because they are not bodies. We say, however, that he means the surfaces, and by saying that they are together he means that they are coinciding. 5

Yaḥyâ:
[226b23-7] 'Between' is that towards which something changing is 13
proceeding before arriving at the goal and the contrary, when the change is continuous and has its natural course. For instance, grey is between white and black, because white changes towards grey before it changes towards black. He says: 'when it has its natural course', because when black becomes grey and then turns back and becomes black again, then grey is not 'between' [black and black], since it is not between contraries. If the change had its natural course, the motion would continue until it reached its goal. If the change is not continuous, there is no 'between' either. For instance, when white changes into grey and the change stops when it has reached grey, so as to cease changing, and then continues to proceed towards black, 20
then grey is not 'between' because, due to its resting in grey, the 540,1
change has become more than one. Change occurs from something – the start – towards something – the goal – and what is in between is that to which the changing object is changing before the goal; therefore 'between' implies at least three things.

Yaḥyâ and Abû Alî:
[226b27-8] Having mentioned continuity in the preceding passage, 9
he will now explain; he says that a continuous motion is a motion which is not interrupted in time, i.e. which is not at rest during any part of the time. He says: 'It is possible that there is an interruption in the place (path) which is traversed by the motion, but in time no interruption is allowed.' For instance, one who is walking does not interrupt his motion in time, although he may interrupt [his contact] with the ground, because the walker may move parallel with a fraction of the ground, so that he does not have contact with all of it.[61] He will specify later what he has said about this, saying that the continuous is what is not interrupted in either time or in place. 15

Yaḥyâ:[62]
[226b29-32] He explains his point that a continuous motion may be 541,5
interrupted in that along which the motion occurs (the path), and yet may not be interrupted in time. He says that one who plays an instrument may pass from plucking the bottom string to plucking the top string, without plucking the strings between these. Then the motion has been interrupted in the path which is traversed by the motion, but it has not been interrupted in time. However, this motion,

having been interrupted in the path which is traversed by it has also been interrupted in time, because some time must have passed between the plucking of the bottom string and the top string; during that time the player was moving from the bottom to the top string. The sound, however, has not been interrupted, because the sound of the top string occurs together with the fading away of the sound of the bottom string.

It is not difficult to show the interruption in the path of a local motion, for instance the motion of a walker. It is difficult for motions which are not local, for it is not easy to explain that they are not interrupted in that in which the motion takes place, although Aristotle says: 'This is evident in the changes which occur in place and in all other changes' [226b31-2].

Yaḥyâ:[63]

542,2 [226b32-4] He has said that a change always proceeds towards a contrary; since it is not quite clear what the contrary of a place is, he intends to explain this. He says that the contrary as regards place is that which is farthest apart, [measured along] a straight line, such as the highest and the lowest place. A straight line is the shortest of all lines which may be drawn between two points, and since it is the shortest line, it is well-determined. The contrary of well-determined is undetermined. Thus the contrary in respect of place is that between which and its contrary is the farthest possible distance, measured along a straight line.

Maybe they want to set forth here the definition of what is contrary in an absolute sense, when they say that what is contrary is such that its distance is the farthest possible distance, with the condition that it is the same kind of thing.

Yaḥyâ and Abû Alî:[64]

543,6 [226b34-227a6] What is 'in succession' to something else is what comes after something of the same kind, either in position, or in form, i.e. in nature, or in any other sense, i.e. in art, in such a way that between the thing and what succeeds it there is nothing of the same kind. An example is our saying that Basra succeeds Kufa,[65] because it comes after it; both are the same kind of thing and between them there is nothing of the same kind, for there is no other town between them, although there are between them other sorts of things, such as rivers, villages and deserts.[66]

Alexander says that by the phrase: ' … .in such a way that there is nothing of the same kind between the things which are in succession' Aristotle means that there is nothing of the same species and nature between them. For a man may succeed another man, although between them there may be something of the same genus, such as an

ant and other [animals]. If there were another man between them
then they would not be in succession.

Another example of things which are in succession is mentioned
by Aristotle, namely that a unit succeeds another unit, and that a 15
line succeeds other lines, and that lines succeed another line, i.e. that
lines as a totality succeed another line, not that each one of them
succeeds the line, unless those lines themselves are ordered one after
another; then they form one row.[67]

As for something succeeding something else in position, we may
posit that, for instance, Basra is first, and then we posit that Kufa 20
succeeds it. One may also do the opposite and posit that Kufa is first,
and that Basra comes after it.

An example of things which are in succession in form or nature is
that the species succeeds the genus; e.g. we say that the crow succeeds
the split-winged birds. In a similar way we may say that things
succeed each other as regards their natural places: as air succeeds
fire and water succeeds air, and earth succeeds water.

By his words '… .or in any other way' [226b35] he possibly intends 544,1
a kind of ordering in art, such as the case of a book in which main
points succeed the introduction and these are succeeded by the
exposition; the exposition, in turn, is succeeded by the criticism, and
that is succeeded by the reply to the criticism. This order cannot be
inverted, nor can the order of things which are in succession by nature
be inverted. The order of things which are in succession as regards
position may be inverted, and one may posit a thing as succeeding, 5
which was before posited as preceding.

As for his words 'What is in succession must succeed something
and must come after something', [227a4] I think that this is an
argument for his proposition that what succeeds must come after
something; therefore the number one does not succeed two, nor does
the night succeed the next night.

Yaḥyâ and Abû 'Alî:
[227a6-7] What is contiguous is specified by all conditions which specify 12
what is in succession, with an extra one. The conditions for what is in
succession are that it comes after what it succeeds, that it is of the same
kind, and that between them there is nothing of the same kind. The
extra condition is that the extremes should be together. It is necessary
that they are of the same kind, for one may say that a man is contiguous 15
with another man, when they are touching, but one does not say that a
stone is contiguous with a man when they are touching.

Yaḥyâ:
[227a7-10] He returns to the discussion of 'between'[68] and says that 545,2
if what is 'between' is at some intermediate position between two

opposites, and if opposites exist in contraries and contradictories, and there is no intermediate between contradictories, then there cannot exist a 'between' in connection with contradictories; it does exist for contraries, but not for all, because there is no intermediate between even and uneven, although they are contraries.

Yaḥyâ:
545,14 [227a10-13] As things which are contiguous are closer to each other than things which are in succession, so continuous things are more specified than contiguous ones. Similarly, things which are in succession are more primary in being, and if their closeness increases, they become contiguous. If they become united, and their extremes become one in actuality, [but remain] two in potentiality, then they are continuous. For the extremes of two continuous, united parts are one in actuality,[69] and are common to both parts, forming a link between them.

As for things which are touching, although there is nothing between them, still they actually exist as two parts, and their extremes are also actually two. In this way we distinguish between what is touching and what is continuous. When the extremes of what is
546,1 continuous are one, then the whole is one, i.e. that which consists of both [continuous] parts is one, and it is actually one, but potentially two. Similarly, the extremes are actually one, but potentially they are two extremes.

[**227a13-17** This definition makes it clear that continuity belongs to things that form a unity naturally by virtue of their contact. And in whatever manner the composite becomes one, the whole will be one, e.g. by rivet or glue or contact or by natural unity.]

Yaḥyâ and Abû 'Alî:
548,8 It is possible for things to be united only if they are of the same kind, not of different kinds. Things of the same kind are for instance the wet and liquid things. Liquid things most properly and primarily admit of becoming united.

By his saying '... by a rivet or glue' he refers to the things which become so close that they are touching and that thereupon their touching becomes being welded. Or 'by a rivet' refers to what is touching and 'by glue' to a being welded, as occurs when someone wants to glue together the broken pieces of a bowl by melting the places where they are touching and then making them into one continuous whole. It is appropriate that after 'being glued and riveted' comes 'being united'.

The order of 'between', 'in succession', 'contiguous' and 'continuous' is that 'in succession' comes first, then contiguous, and then continuous. For nature makes being-in-succession primary, because the units (numbers) have the property that they can be in succession, but they cannot be touching, because they do not have a [spatial] position and therefore are by nature primary in relation to what does have a position. The separate units which do not have a position count[70] the continuous things which do have a position, and therefore are prior to them by nature.

Continuity comes last in being [of the above-mentioned relations], for it arises from the contiguous or from what is in succession, because 549,1 continuity must originate from what is not continuous, and it necessarily arises[71] only from what is not continuous, i.e. the contiguous and what is touching. As there is no touching between units, the continuous cannot originate from them.

As there is[72] no touching between units, nor can there be, whereas between points touching is possible, point and unit are not the same thing, as the Pythagoreans assert, saying that a point is a unit with a position, whereas a unit is a point without position. Furthermore, it is impossible that there exists something between two [adjacent] units, whereas between each pair of points a line may exist.

By his remark[73] that points are touching he does not mean that they are properly touching, because things which are touching touch each other in their extremes and limits, and points do not have limits and extremes. By 'touching' he means here being united (coinciding).

The continuous is that of which the parts have one common limit.

In his statement[74] '... in which things they exist and in which ones they do not exist' he refers to the concepts 'in succession' and 'between'; he has explained that they exist in things which have a [spatial] position and in things which do not have a position. The other concepts exist only in things which have a position.

[Aristotle 5.4]

{227b3-14 Changes[k] may be one (a) in genus, (b) in species or (c) in an absolute sense.}

Yaḥyâ:[75]

Some concepts which were defined [in the preceding chapter] are used 551,8 in an explanation of the statement that the continuous cannot be composed of indivisible parts; e.g. a line is not composed of points, a time not of instants and a motion not of *motions*,[76] i.e. starts of motions. Also, some concepts defined above are used to explain what a change[k] is which is properly one. A change[k] is one in the same variety of ways as the word 'one' is used. 'One' may be one in genus, or one in species, or one in number. Change[k] may be said to be one in any of these ways.

A change[k][77] is one in genus if it falls under one of the other genera.[78] Two examples of local motion are: circular motion and rectilinear motion; both fall under change of place. Examples of change[k] falling under the category of quality include becoming white and becoming black; both are changes[k] in quality. A further example, regarding change[k] in the category of quantity, is growth and diminution. There is no single category under which all changes[k] fall, because quantity, quality and the 'where' are higher genera, which do not have a common genus under which they might fall, since what exists cannot be considered to fall under one common genus which comprises everything. If all changes[k] do not fall under one genus, then 552,1 the word 'change[k]' is a homonymous word, not a synonymous one.

Changes[k] which are one in species are those which fall under the same indivisible species. An example is one process of becoming-white and another. Becoming white and becoming black cannot be said to be one in species, because although they both fall under becoming coloured and colour is a species of quality, colour is a genus for both white and black, so one cannot say that it is a species [in the required sense]. Therefore the changes[k] which fall under it are not one in species at all.

For a change[k] to be one in number (individually one) three conditions must be fulfilled: that its subject is one, that its time is one, and that the species and the form at which the change[k] ends is one. An example is the case in which I become healthy at a certain specific time. This is a change[k] which is one in number, because its subject is one individual, namely me, and also the time is one individual time, and the kind of change[k] is of one individual kind. If I become healthy and change place at one and the same time or I become healthy and you change place at the same time, or I become healthy and become healthy again at two [different] times, then in each of these cases the changes[k] are not individually one.

Among the conditions for a change[k] to be one in number, he does not mention that the efficient cause (mover) must be one or that the starting point of the change[k] must be one, because if the three conditions he has mentioned are fulfilled then these two conditions will also be fulfilled; they necessarily follow from the three mentioned conditions.

227b11 When it happens that something is a genus and a species at the same time, then it is clear that a change[k] will be one in species, but it will not be a change[k] which is one in species in an absolute sense.

Yaḥyâ:
Of [changes[k] in] the intermediate genera we cannot say in an absolute

sense that they are one in genus because they fall under the [same] intermediate genus, nor can we say that they are one in species, because that in which the change[k] occurs is not a genus in an absolute sense, nor a species in an absolute sense. An example is learning; this is a change[k] towards knowledge and knowledge is a species of apprehension,[79] and it is a genus of the particular kinds of knowledge, so it is an intermediate genus. Therefore of learning one cannot say in an absolute sense that it is one in species nor that it is one in genus. 553,1

> [**227b14-17** There may be a puzzle about whether a certain change[k] is one in species when the same thing changes[k] from the same to the same; for instance, when a point changes[k] from this place to this place again and again.]

Yaḥyâ:
554,10 The problem is of the following kind: If change[k] is one in species when its subject is one and the same, the starting point is one and the same and the end point is one and the same, then it necessarily follows that if someone covers the same distance twice, one time along a straight line, the other time along a winding path, these two motions are one in species, i.e. the rectilinear motion and the winding motion.

The solution is that the manner of the motion should be the same, and rectilinear motion, circular motion and winding motion are not of the same manner.

One should note[80] that if the goal of the motion is *not* one, then the manner of the motion to it may <not> be one. An example is the case in which something covers one and the same distance one time along a straight line and another time along an winding path. It is also possible that it covers the same distance with two rectilinear motions; then they are of the same manner.

By the phrase 'If that in which the motion occurs is different in species, then the motion is different' [227b19] he means that if the path along which the motion occurs is circular or winding, then this 555,1 path is different in species from a rectilinear one, along which a motion may take place. And if the paths are different in species then the motion along them is also different in species.

> **227b21** A motion is said to be one in an absolute sense if it is one essentially and numerically.

Yaḥyâ:
The expression 'in an absolute sense', stands for 'in the true (proper) sense'. A motion which is one in the proper sense is a motion which is essentially one, and a motion which is essentially one is a motion which is one in number. Such a motion[81] is most properly one, because

it takes place in things which exist by themselves. A motion which is one in species or in genus is only known in the mind.

> **227b22** What such a motion is will become clear from making distinctions.

Yaḥyâ:
This means: by distinguishing the things in which motion takes place, such as time, place and so on.

> **227b23** There are three things about which we speak in connection with motion.

Yaḥyâ:
In an earlier passage he has enumerated five things:[82] that which is in motion, the cause of the motion, that from which the motion starts, that towards which the motion proceeds and the time. Now he enumerates three things because they are more proper and useful in the course of his discussion. The first thing is that which is in motion, the second one is the kind of thing in which the motion occurs and the third one is the time. We say that if these three things together are one, then the motion is one in number.

Yaḥyâ and Abû 'Alî:
556,16 [227b27-228a6] Having mentioned the three conditions which must
557,1 be fulfilled together for a motion to be properly one, he now makes clear that each condition taken separately does not make the motion properly one. He says that if the motion is in one thing, i.e. in one kind of thing, and if the subject of the motion is different or the time in which the motion takes place is different, then the motion is not properly one, but [only] one in species or genus.

If the time of the motion is one and continuous and the kind of motion is different, then the motion consists of successive motions. For example, suppose we have a continuous period of time, and in a part of it Zayd becomes healthy, and then he stops being healthy and becomes ill again in the rest of the time; then his becoming ill follows after he stops being healthy and we have two successive motions, no matter whether they are of the same species, or of two [different] species.

> **227b31** What is in motion must be one, not accidentally, like the black which becomes white[83] and Socrates[84] who walks.

Yaḥyâ:[85]
He says that the subject of the motion must be one in order that,

together with the two other conditions, it will be true that the motion is one. It is not sufficient that the subject is one in an accidental sense, for if Socrates is learning and becoming hot during the same time then these motions are not properly one, although the subject of both motions is one, the time is one and both motions fall under [the same genus, i.e.] quality. The reason is that the subject of the motions is not properly and essentially one, because the subject of heat is the cold,[86] but the subject of learning is not the cold; it is accidental for Socrates, who is the subject of learning, that he is cold, so that he becomes a subject which may accept heat.

Yaḥyâ:
We must give the example in this way, i.e. that Socrates is *walking,* learning and becoming hot, or other such motions which fall under quality. This is a more appropriate example than the one given by Aristotle, namely that the black becomes white and that Socrates is walking, because walking is a local motion, and becoming white is a change[k] in quality. This is already sufficient for the changes[k] not to 558,1
be one, and it is not necessary to say that it is not one on account of its subject being one only accidentally.

> **228a1** ... nor [because the subject is] common [in nature], because it is possible for two men to be subject to the same process of getting healthy.

Yaḥyâ:
He says that if the subject is one essentially, not accidentally, then it is [not only] necessary that it is common [in nature], such as when one says that Zayd becomes white and ʻAmr becomes white, but [for a motion to be one in number] it is also necessary that the subject is one in number, not [only] common [in nature].

> **228a3** Suppose that Socrates undergoes an alteration which is one in species, but at one time and then again at another time.

Yaḥyâ:
He says that although the time is one and continuous the motion might be not one.[87] An example is that Zayd becomes healthy in one continuous time, first at one time and then at another time. Then the first healthiness is not the same as the second one, except in species, because the first one perishes and is followed by some illness. When the second healthiness comes into being after another illness then it cannot properly be the same thing as the healthiness which has perished, but it is the same only in species. If both conditions of healthiness are not one in an absolute sense, then the changes[k]

towards them, i.e. the processes of becoming healthy, are not one in an absolute sense either.

{228a6-20 Digression: under what conditions is the form towards which a change proceeds numerically one?}

Yaḥyâ:

559,16 Having mentioned the problem whether motions towards health occurring in two different times are one and the same on account of the fact that the species of motion is one and the subject is one, such as when Zayd becomes healthy, then falls ill, and then becomes
560,1 healthy again, Aristotle transfers the problem to the form towards which the motion proceeds. The form must be one and the same if the subject is one, even if the time is different. An example is the health of Zayd in the morning and his health in the evening, when there is a period of illness between them. The problem becomes more acute when we say that if the health had lasted from morning until evening, without a period of illness in between, or if the process of becoming healthy had lasted from morning until afternoon without interruption, the motion of becoming healthy would be properly one. Similarly the health itself, which we have supposed to exist without interruption, would be one, in spite of the fact that the subject is flowing and not stable. If it is flowing and not stable, then a condition like health and whiteness is not numerically one. How then could it be said[88] that the health is numerically one even if there is a period of illness in between? Aristotle has not solved the problem.

The solution is as follows: we say that whiteness or health is one and the same in the proper sense, if the time is one and the same and the subject is one, in the sense that besides the substrate[89] being one, the form, i.e. the white, has not perished to become black, but to remain white. It may increase or diminish with the flowing [of the substrate of the white]. Our statement on perfection (*at-tamâm*) that it is numerically one, is like our saying of Zayd that he is numerically one.

Alexander has solved the problem, saying that although the substrate is in flux, in any case something of it remains stable, and what has a condition resembles it so that [one may say that] that it is numerically one.[90]

228a12[91] The difference, however, is that if the activities are two, then from this very fact it follows that the states are also like them [two] in number.

Yaḥyâ:

Having mentioned the question under what conditions motions which proceed to the same form are one and the same and the question under what conditions a form is one and the same, Aristotle says that he now will mention a certain difference between these problems; in connection with this difference he mentions the motions which proceed from states, whereas before he has been speaking about motions which proceed towards states. What he says about both [kinds of] motion is the same thing.

The difference is the following: if the activity proceeding from a state – i.e. the motion – is one, then the state from which the activity proceeds is also one. If the activity proceeding from a state, i.e. the motion, is not numerically one, then the state is not necessarily many, but still *may be* numerically one, whereas on the other hand, if the states are many, the activities are necessarily many, not numerically 561,1
one; for from one potentiality many activities may arise. Thus a plurality of activities does not require a plurality of states from which they proceed, but a plurality of states necessarily requires a plurality of activities.

The discussion of the motions proceeding towards a form is similar: a plurality of motions does not necessarily require a plurality of forms, whereas a plurality of forms necessarily requires a plurality of motions proceeding towards them. For we know that the two angles which stand on the base [of an isosceles triangle] are equal, both by direct proof and by a proof *per impossibile* (*bil-istiqâma wal-k̲ulf*). If the process is one, then the form is necessarily one, and a plurality of forms requires a plurality of motions proceeding towards them. A plurality of motions does not require a plurality of forms.

228a7 Is health one, and in general are the states and affections in bodies essentially one?

Yaḥyâ:
By the expression 'states' he means objects of knowledge.

By 'affections' he means the corporeal qualities, which are called states. He says 'essentially' instead of 'in number'.[92]

228a9 For we find that what contains them is in motion and flowing.

Yaḥyâ:
He explains something which adds to the problem: if the state is numerically one and decreases in conformity with the subject, why would the state not be numerically one, even if it exists at two different times, i.e. the state which has perished and the one which has come to be after the perishing?

228a12: For the argument is the same.

Yaḥyâ:
He says that the problem concerning the states is the same as the one laid down for the motions and it can be compared to it, namely by
25 saying that if the state is numerically one, while the subject is flowing, then why would the state not be numerically one, even if it includes a perishing?

228a12 The difference, however, is ...

Yaḥyâ:
562,1 Having mentioned the problem concerning the states being numerically one and the problem concerning the motions being numerically one and having discussed these problems, he now explains to what extent these problems are different.

228a18 It would be possible that something which is one and the same perishes and again comes to be many times.

5 Yaḥyâ:
He does <not>[93] solve this problem; he says that if a motion which is numerically one were to include a rest, then it would be possible for the same thing to come to be and perish many times, which is absurd. We should know that the inquiry about states is outside of what we are discussing at present.

{228a20-229a6 In order for a motion to be numerically one it must also be continuous, complete and uniform.}

564,10 Yaḥyâ and Abû 'Alî:
Having discussed the above-mentioned problems, he returns to the subject under discussion, namely the question of what it is for a motion to be one. He explains this by a long passage in which he says that a motion is one and continuous when three conditions are fulfilled: that the time of the motion is continuous, that its subject is one, and that the species of motion is one. If every motion is divisible,
15 every motion is continuous, because the definition of the continuous is that it is divisible into parts which are always further divisible. If every motion is continuous, and the continuous is that of which the parts have extremes which are one and common, then a motion which is properly one must be continuous. For if a motion is numerically one, then its time is numerically one, its species is one, and its subject is one, and this is possible only if the motion is a unity and its parts

have extremes which are one. If every motion which is numerically one is continuous in this way, then similarly every continuous motion is properly one, because it is only continuous if the subject, the time and the species are one.

The extremes of any two arbitrarily chosen things generally do not coincide or form a unity; this is only possible for things which are of the same species. Therefore the end of a line and the end of walking cannot be one. Thus motions which are different [in species] can be neither one nor continuous, even if their times are continuous. They are contiguous, as when a man is running and immediately after his running, catches a fever.

His statement: 'Since every motion is divisible ... ' [228a21] is a proof for his claim at the beginning: 'Every motion is continuous' [228a20].

He said: 'Some things have no extremes at all, and others do have extremes, but their extremes are <different>[94] in form, and are called extremes by homonymity' [228a24]. He says this because having mentioned the extremes in general, he wants to elaborate on them and explain which things have extremes and which ones do not have them. Extremes occur only in magnitudes. If extremes occur in other things, then such things are accidents of magnitudes, because extremes belong to things which are continuous, and the continuous is essentially a magnitude, not a motion. Motions have a continuity which is derived from [the continuity in] a magnitude. Some things, such as points, have no extremes at all, because a point is an end [itself] and is not bounded by an end.

Divine things[95] do not have extremes either; therefore they are not continuous. The things which are subject to generation do have extremes, but only those among them which are of the same species have extremes which may form a unity and become one. Things which are of different species have extremes which [only] have the name 'extreme' in common (which are called 'extreme' homonymously), because they are of a different species, such as a line and getting healthy. These and similar things cannot be continuous.

> **228a28** For example, the torch-race is a contiguous, but not a continuous locomotion.

Yaḥyâ:[96]

He says that if two men each have a torch, and each of them quickly turns over the torch he has in his hand to the other, and each of them tries to be quicker than the other in doing this, then the motions of these men are not continuous, because the subject of these motions, i.e. the two men, is not continuous. The motions are not even contiguous, because a stopping of the motion necessarily occurs at the instant the men turn over the torches. If that is the case the motions of the torches are not contiguous, but successive.

> **228a30** Their time is continuous on account of the continuity of the motions.[97]

Yaḥyâ:
He means that the motion is continuous because its parts are continuous with one another and between each pair of parts there is one common limit.

Yaḥyâ and Abû 'Alî:[98]
567,5 Having explained that a motion is only numerically one if its subject, its time and its species are one, and having stated that what he has mentioned is what he intended to mention, he adds two more conditions. One condition is that the motion must be complete, i.e. it must proceed towards a complete form without stopping on the way, as becoming white proceeds to the form of whiteness without stopping on the way and then returning. For just as we do not say of someone who has not completely become a man that he is one man, we do not say about a motion which is not complete that it is one motion. How could it be one, when it is not complete and finished?

The other condition is that motion must be uniform, not irregular. It may be irregular either because of something in the motion itself, or because of something in the subject along which the motion occurs. An example is a motion along a bent line [sc. one without *constant* curvature]; such a motion has irregular parts, because the subject along which it takes place is irregular in its parts; i.e. the parts cannot be brought to coincidence with one another. This does not hold for a straight line and a circular line, because the parts of such lines may be made to coincide with one another.

A motion which is irregular because of something in itself, not its subject, is a motion which occurs along a straight line, but which is partly slow and partly quick. This may occur when the road is long. Then what moves may move slowly at times and quickly at other times. Then the motion is irregular and divisible, i.e. it may be divided in a slow part and a quick part, and it is not one, for if we conceive its parts in our mind we find that they are not similar. A uniform motion comes to be if it occurs along a straight line and if its parts are regular, i.e. do not differ in quickness and slowness. This is the way in which the [celestial] bodies[99] move; they are moving with a complete motion, because they always admit the intellect.[100] The
568,1 motions in our world are not so strictly rectangular and circular and are not from proper oppositions. Therefore they do not proceed to a

perfect form, for the process of becoming white does not go from a purely black colour to a pure white.

228b11 A motion is also said to be one when it is complete, no matter whether it is complete[101] in species or in essence.

Yaḥyâ:
He means that if a motion is one in number, in species or in genus, it must be complete in order to be said to be one in number or in species. An example of motions which are one in species are two processes of becoming white in two different subjects. They must be complete in order to be said to be one in species, for one does not say of what is not complete that it is that thing at all. Therefore one does not say about a generation that it is a generation when it is not proceeding towards a complete form, but then one says it is only half a generation.

228b14 Sometimes a motion is called one, even if it is not complete, provided only that it is continuous.

Yaḥyâ:[102]
If a motion is not complete, but continuous, then it is one because of its continuity, not properly one. This occurs when the process of becoming white stops before it has reached the complete form of the white. Then we say that the becoming white, until it stops, is one motion, because of its continuity until the instant of its stopping. It is not one in an absolute sense, because it is the goal of nature to arrive at this form by the motion. If the goal is not reached, the motion is not one.

228b18 The difference seems to be a question of more or less [degree].

Yaḥyâ:
He says that uniform and non-uniform motions are not different because of the fact that uniform motion is one and non-uniform motion is not one, but because the latter has less unity and the former has more unity. Thus, non-uniform motion is less one and uniform motion is more one.

Yaḥyâ and Abû 'Alî:
It is necessary that a motion is either complete or not. Similarly it is necessary that it is either uniform or not. 570,14
Uniformity of a motion may occur on account of that along which it takes place, or on account of the motion itself, because of the fact that the whole motion is quick

or the whole motion is slow. A motion is either slow or quick or partly quick and partly slow. Slowness and quickness are not species for a motion, nor specific differences.

Yaḥyâ [ad 228b30]:[103]

571,16 He does not mean heaviness and lightness in an absolute sense, because those are specific differences which are said of bodies. He means, however, heaviness and lightness in a relative sense, such as the heaviness of a piece of earth and the heaviness of another which is heavier than the former. Then we say that the former is light and the latter is heavy, whereas both tend to move to the centre. Such heaviness and lightness are not specific differences, because they exist in all bodies which move towards the centre. It is just as one does not say that for a large piece of fire and a small one <large and small are specific differences>.[104]

572,5 **228b25** Sometimes [the irregularity] exists not in the thing,[105] nor in the time, nor in the goal, but in the manner of the motion.

Yaḥyâ:

By 'not in the thing' he means: not in the magnitude along which the motion occurs. By 'nor in the time' he means that the cause of the irregularity of the motion is not the difference of two times. By 'nor in the goal' he refers to the form, i.e. the irregularity and the lack of uniformity of the motion do not arise because of an irregularity in the form towards which the motion proceeds.

The irregularity[106] occurs because of the motion itself, as when it is partly quick and partly slow, because of an irregularity in the force. We do not say that the irregularity of such a motion is due to the time, the time of the slow motion being longer than the time of the quick one. For although this is indeed true, the irregularity arises on account of the force.

228a30 Nor are heaviness and lightness, when they refer to the same subject, such as one piece of earth and another or one piece of fire and another, a species or a specific difference for motion.

Yaḥyâ:

He says this in order that one should not think that by 'heaviness' he means the heaviness existing in earth and by 'lightness' the lightness existing in fire, but by the expressions 'heaviness' and 'lightness' he means that one piece of earth is compared to another, which is lighter or heavier, which more properly amounts to the same thing.[107]

> **229a2** A lesser degree of something always arises by an admixture of its contrary.

Yaḥyâ:[108]
He says this in order that nobody will say that if an irregular motion has less unity than a uniform motion, it cannot be one at all. If the 573,1
motion is continuous, then he says there is no objection to the claim that it is one to a lesser degree than a uniform motion is said to be one, because a mixture of two contraries may give rise to a lesser degree of something. Therefore an irregular motion may be one only in an imperfect sense, namely when it is continuous.

A lesser degree of something arises by an admixture of its contrary, as the imperfectly white arises through a mixture with black. In this way an irregular motion may be one to a lesser degree; when this is the case it must have arisen from an admixture with plurality.

> **229a3** Since every motion which is one can be uniform and non-uniform, contiguous motions which are not of the same species cannot be one motion.

Yaḥyâ:[109]
He says that since every motion which is one can be uniform one time and non-uniform another time, such as the motion which sometimes takes place along a straight line and at other times along a line which is not straight, but curved, and since a uniform motion is homeomerous (i.e. a motion of which any arbitrarily chosen part may be brought to coincide with any other part) it necessarily follows that contiguous motions cannot be one if they are different in species, such as walking and learning. For if they were one, it would be possible that sometimes they would be uniform, which implies that one motion would coincide with the other, and this is absurd.

[Aristotle 5.5]

> {229a7-229b22 Where F and G are contraries, the movement *from* F to G is contrary to the movement *to* F from G.}

Yaḥyâ and Abû Alî:
His aim is to perform an investigation in order to know which motions 579,11
are contrary to each other, which rests are contrary to each other and which motion is contrary to which rest.

He first investigates which motions are contrary to each other. Contrary motions should be looked for among motions which take place in contrary things. The contrary motions must be related either

to one contrary or to two contraries. In the case in which they are related to one contrary, an example is the motion from black and the motion towards black. When a motion is related to a contrary, it is well-known that it has a beginning and a goal. Then both [contrary] motions must be related to the contraries[110] by one of these – i.e. the beginning or the goal – or by both of these together. If they are related
580,1 to them by one of these, then either by 'towards' or by 'from'. If the relation to them is by 'towards', then an example is the motion towards black and the motion towards white. If they are related to them by the relation 'from', i.e. by the beginning, then an example is the motion from black and the motion from white. If they are related to them by 'from' as well as by 'towards', then either they are related to both of them by 'from' and 'towards', or to one of them by 'from' and to the other by 'towards'. An example of the latter case is the motion towards black and the motion from white. If they are related to both contraries, then the relation is the beginning and the goal, i.e. 'from' and 'towards'. An example is the motion from black to white and the motion from white to black.

Not all these subdivisions of motions are necessary. The motions from black and towards white are one and the same in essence; they only differ in relation. i.e. one is related to white and the other to black. Since these motions are essentially one, it follows that if they were to be contrary, the motions would be contrary to themselves.

Change includes motions and other kinds of change; motion is change from one contrary to another; change towards something is generation and change from something is perishing; both are changes: the latter is from a subject and the former is towards a subject; they are not changes from one subject to another – change from one subject to another is always a motion, nothing else. Therefore it is necessary that a contrary motion is from one subject to another, i.e. a motion in which there occurs a change from a subject to some form. Thus motions which are contrary are, for example, those of which one is from black to white and the other from white to black, because the concept of motion implies that it is from something towards something.

Consider two motions, one of which starts from one contrary and the other from the other contrary, or two motions, one of which is from one of the contraries and the other towards it, such as the motion from black and the motion towards black; one does not find in each of those motions a 'from' as well as a 'towards', and therefore they cannot be called motions in the proper sense, so the proposition that they are contrary is not true.

581,1 Having said[111] that the motions which are related to the contraries by 'from' and the motions which are related to the contraries by 'towards' are not contrary, Aristotle says that those which are related to the contraries by 'towards' have a better claim to be called contrary than those which are related to the contraries by 'from'. For although a motion is always from something towards something, it is related to the goal towards which it proceeds, and it is the goal from which it derives its name, not from the starting-point. The motion from illness to health, for instance, is called becoming healthy, not becoming ill. Motion towards black from white is called blackening, not whitening. If a motion were not better characterised by its goal than by its starting-point, its relation to the 'from' should not be severed and [only] its relation to the 'towards' established. How could it not be like this when they are one in invalidating the starting-point?

229a22 This we shall discuss later.

Yaḥyâ:[112]
He will later discuss the question of [change to] an intermediate and explain how a change is not only from a contrary and to a contrary, but may also occur from an intermediate state and to an intermediate state.

229b2 From a consideration of examples (induction) it becomes clear which processes are considered to be contraries. For we see that the contrary of someone becoming healthy is that he becomes ill.

Yaḥyâ:
When someone is becoming ill, this means that he changes from health to illness; his becoming healthy is the contrary of his becoming ill.

229b4: Being taught is contrary to being led into error but not by oneself.

Yaḥyâ:[113]
By the expression 'being taught' he means that someone changes from an incorrect opinion to a true one and that this is brought about by a teacher. For our expression 'being taught' implies that the knowledge is acquired from someone else.

By the expression 'being led into error' he means that someone turns from a true opinion to another one. He adds the expression 'not by oneself' in order that it becomes opposite and contrary to 'being taught'. Since being taught is due to someone else, being led into error, which is contrary to it, must also be due to someone else. Being led into error from having no knowledge or opinion at all is not a motion, but a change, and turning from having no opinion to a true 582,1 opinion is also a change, not a motion. Therefore by 'being taught' and 'being led into error' he must have meant going from one thing to another, i.e. from one opinion to another.

> **229b11** In all cases of something that has no contrary the processes from it and towards it are contrary changes.

Yaḥyâ:
He explains that going towards black and moving from black are not contrary motions, because if there is only one contrary, i.e. black, and going to it and from it are regarded as contrary, then these processes are not motions, but changes only. He compares them to generation and perishing, because what characterises generation is the goal, and what characterises perishing is the starting-point.

> **229b14** Changes[k] to a state between opposites – whenever opposites have intermediate states – must be regarded in a sense as changes[k] to one of the contraries.

Yaḥyâ:[114]
The intermediates between contrary extremes behave as the contrary extremes in the respect that change[k] may occur from them and towards them: a change[k] is possible from black to grey and from white to grey, and also from grey to black and to white. If it occurs from grey to white, it occurs from grey insofar as there is some black in it, not insofar as there is some white in it, because a change[k] does not occur from something to itself; in relation to white grey plays the part of black and in relation to black it plays the part of white.

[Aristotle 5.6]

Yaḥyâ and Abû 'Alî:[115]
586,2 [229b23-230a18] Having explained above which motion is contrary to a motion, he now explains which rest is contrary to a motion. For rests may be contrary to each other, motion may be contrary to rest and motions may be contrary to each other. The contrary of motion is rest because motion is a perfection, a perfection is something actually existing, and rest is the privation of motion; what is a privation of something is contrary to it. It is, however, a contrary in a certain sense, not in an absolute sense. Therefore rest is a contrary of motion in the sense in which a privation is called a contrary. For privation and form cannot exist together in one subject. They are called contraries from this point of view.

Yaḥyâ:
One should be aware that in his discussion above he has said that the form and the privation are contraries and it is understood that there he takes privation not in the sense of what is potential, as he has done in the *Categories*,[116] but in the sense of a privation which is actual. For the privation which is actual may return to the form; the privation which is potential does not return to the form.

229b25 The privation may be called a contrary in a certain sense: a motion of a certain kind has the privation of that kind [as a contrary]; for instance the contrary of local motion is rest in a certain place.

Yaḥyâ:[117]
He says that not just any arbitrary privation, i.e. any arbitrary rest, can be a contrary of any arbitrary motion, but only rest and motion of the same genus can be contraries. For contraries necessarily fall under the same genus. Rest in a certain place, for instance, may be contrary to local motion, not to motion in quality or quantity.

229b27 This statement, however, is said absolutely (without qualification).

Yaḥyâ:
He means that his statement that rest in a certain place is contrary to local motion is not absolute, because rest in a place is not contrary 587,1
to any local motion, but rest in a certain place is contrary to a certain local motion.

229b29 We say that if motion occurs, it takes place between two subjects.

Yaḥyâ:
He clarifies the question he has posed, saying that the motion occurs between two subjects, namely the starting-point and the goal. By 'subject' here he does not refer to the moving thing, which is the subject of the motion. Then he says that natural motion has as a contrary a non-natural motion, which has the reverse direction; for instance, the downwards motion of a stone is natural for it, and it is contrary to an upwards motion of the stone, and this is its only contrary. A motion of the stone to the right and to the left are not contrary to its downwards motion, nor to its upwards motion.

229b31 At the same time these two are also contrary to one another.

Yaḥyâ:[118]

15 He means that the kinds of these [motions] are contrary, i.e. one is natural and the other is non-natural, for example an upward motion of a stone and its downward motion.

> **229b32** For it would be absurd to suppose that the motions are contrary, whereas the rests are not be contrary to one another.

Yaḥyâ:

20 He says that every motion exists on account of its perfection, because whitening exists on account of the white, since by means of the process of whitening something becomes white. Similarly the contrary of a motion exists on account of its perfection. All rectilinear motions have a perfection, otherwise they would exist in vain. Since a motion exists on account of its perfection, then if the motion has a contrary, it is most proper that the contrary exists because of that from which it derives its existence, i.e. its perfection.

588,1 > **230a2** For example, a state of rest in health is contrary to a state of rest in illness, and also contrary to motion from health to illness.

Yaḥyâ and Abû 'Alî:

He denies that motion towards health is contrary to a state of rest in health, because it is absurd that something should be contrary to that towards which it proceeds.

8 > **230a6** It is necessary that it is contrary to this or to that; for it is not possible that resting in white is the contrary of resting in health.

Yaḥyâ:

He says that a motion can only be contrary to a rest when it is either the rest from which the motion started, or the rest to which the motion proceeds. The motion is not contrary to the rest to which it proceeds, because that is its perfection; thus it must be contrary to the rest from which it started. A motion can only be contrary to these two states of

15 rest, because it cannot be contrary to a rest of some other genus, as contraries must always be of the same genus. Therefore a state of rest in white cannot be contrary to a state of rest at a place.

21 > **230a7** As for the things which do not have contraries, they have opposite changes; i.e. the change which occurs from the thing is contrary to the change which proceeds towards it. These, how-

ever, are not motions. So there is no remaining in those things either.

Yaḥyâ and Abû ʻAlî:
The change in the cases of generation and corruption does not occur 589,1
from one form to another; therefore there is no motion in them. If there is no motion in them, there cannot be contrary motions, and if there is no motion in them, there cannot be a state of remaining and rest in them either. For rest is contrary to motion and there is no motion in them.

Yaḥyâ and Abû ʻAlî:
Matter is non-existing, i.e. it is not a particular something, because it does not have any form; therefore a change occurring in it towards a form is a change from not-being to being, and a change from a form back to matter is a change from being to not-being. The absence of change in matter is the absence of change in not-being and the absence of change in the form is absence of change in being. The absence of change in the form and its absence in matter are not contraries. They are contraries when they occur in opposite forms. If matter were a substance with a form, then it would necessarily follow that the absence of change in it would be contrary to the absence of change in a form which has arrived in it.

Aristotle says that one might ask whether the absence of change in matter is contrary to something, when it is not contrary to the absence of change in something existing nor opposite to anything else. He brings up another difficulty,[119] solves it and then turns to the solution of this difficulty.

230a14 If this is the case, then either not every state of rest is contrary to a motion, or generation and perishing are motions.

Yaḥyâ and Abû ʻAlî:
He says that if the absence of change in something is rest, then it necessarily follows either that there is no motion contrary to that rest, so that the proposition that every rest has a motion as its contrary is not true, or that there is a contrary motion, so that perishing is a motion.

230a15[120] It is clear, however, that one should not call this [absence of change] a rest, but one should call it similar to rest.

Yaḥyâ:
He means that the absence of change in something should not be called rest, but should be called similar to rest.

590,1 **230a17** Either it will not be contrary to anything, or it will be contrary to the perishing.[121]

Yaḥyâ:
He means that either the absence of change in something is contrary to the thing's perishing, or it has no contrary at all, or it is contrary to the absence of change in the thing's not-being.

230a17 For perishing is change from it.

Yaḥyâ:[122]
The perishing occurs from the absence of change in the thing.

230a18 Generation is change towards it.

He means that the generation proceeds towards the absence of change in the thing.[123]

{230a18-b21 Natural motion is contrary both to non-natural motion and to rest.}

Yaḥyâ and Abû 'Alî:
593,14 He brings up the following problem: How is it that only local motions have the characteristic that for the same thing one motion is natural and another one non-natural? For instance, the upward motion of fire is natural and its downward motion is non-natural, and the downward motion of earth is natural and its upward motion is non-natural. For a motion like growth and diminution this is not so: neither of them is non-natural. The same holds for alteration, for neither whitening nor blackening is non-natural. Also, neither becoming
594,1 healthy nor becoming ill is non-natural, but both are natural.

By 'natural'[124] we do not mean here that which becomes perfect, for becoming ill is not becoming perfect. By our saying that becoming ill is natural we mean that it does not occur through an external cause or by compulsion. If things are not eternal their forces are finite and therefore these forces diminish and are cut off with the passage of time. Thus becoming ill occurs in a natural way, because it does not occur by compulsion or some external cause.

Similarly, also in the case of generation and perishing the change is never non-natural. He solves the problem by saying that among all these motions[125] there are those which are non-natural and those which are not.

Growth and diminution may be natural, i.e. when something grows in its proper time and to the size which is specific for it, not more, and similarly for decrease. Something may grow more and become larger than its proper size, or it may not grow in its proper time, because of a surplus of food or a surplus of heat, such as seeds which grow before their proper time and become larger in size than they should, so that they get a ridiculous size (*bi-an yaskufa ḥajmihî*). This is non-natural. A non-natural diminution is, for instance, the cutting off of a limb.

Similarly, becoming ill is a natural process if it is not caused by something external or by compulsion. It is a non-natural process when it occurs because of an external cause or bad food. In the same way becoming healthy is non-natural if it leads to health on a day which is not the critical day (*yawm buḥrân*); if it occurs on the critical day the process is natural.[126] Similarly someone who becomes dark by the heat of the sun and someone who becomes hot by an excessive motion – these are all non-natural processes.

The same[127] applies for change in the case of generation and perishing, for perishing by a murder is non-natural; perishing which does not occur by compulsion, but by a diminishing of forces is natural. Generation occurring at its proper time in the way it should occur is natural, but it is non-natural in other cases, as when a boy rapidly attains puberty because of a luxurious life full of pleasures, or such as the coming to be of fruits and their seeds because of an excessive heat. All these non-natural things are not praiseworthy.

Then he brings up[128] another problem, as follows: if some processes of perishing are natural and others non-natural, then one perishing may be contrary to another. But generation is also contrary to perishing, so one thing would have two contraries, which is impossible.

The solution is that it is absurd that one thing should have two contraries from the same point of view, so that white as white would have black as a contrary and also something else. But it is not absurd that something has one thing as a contrary from a certain point of view, and something else as a contrary from another point of view. Then from that point of view it has only one contrary. For example, courage is contrary to cowardice and to recklessness, insofar as courage is a virtue and both others are vices. Thus virtue is contrary to one thing only, namely vice. Recklessness and cowardice are not contrary to each other as virtue and vice, but as things which are something to a greater and a lesser extent. Cowardice is contrary to recklessness in this respect, and it is contrary to courage as vice and virtue. Thus vice is contrary only to virtue and a greater extent is only contrary to a lesser extent and nothing is contrary to more than one other thing.

Similarly, a natural illness is contrary to health insofar as they are health and illness and it is contrary to an illness which is non-natural not because they are both illnesses, but because one is natural and the other is non-natural. Illness as illness is contrary only to health, and the same holds for becoming ill and becoming healthy. A natural process of becoming ill *qua* natural is contrary only to a process of becoming ill which is non-natural. Thus something is contrary to one other thing only.

230a21 An example is alteration; it is not the case that some alterations are natural and others non-natural. For if someone becomes healthy or falls ill, then one of these processes is no more natural than the other.

Yaḥyâ:
Becoming ill and becoming healthy are not alterations in the proper sense, because the change[k] of alteration does not fall under [changes to] privations and forms, i.e. generation and perishing, and in general no change[k] falls under this at all.[129] He has clearly explained this above, when he set forth that changes[k] do not fall under the whole category of quality. If becoming healthy is a change from a non-subject to a subject and becoming ill is a change from a subject to a non-subject, then they are not changes[k], and therefore they are not alterations in the proper sense. He applies the word to them because he has not explained these matters before.

230a31 Are there are also kinds of generation which occur by compulsion, not in the naturally ordained way (*heimarmenôs – muqaddiran taqdîran*)?

Yaḥyâ:[130]
596,1 By the expression 'in the naturally ordained way' he means 'by the order and course of nature'.

230b6 Then two perishings are contrary to each other.

Yaḥyâ:[131]
This is the second problem, namely that if the contrary of a perishing is another perishing and also a generation, then there are two things contrary to one thing.

230b7 There is nothing that prevents this.

Yaḥyâ:[132]
He says that there is no objection to one thing having two contraries. His phrase: ' … .if one perishing has this quality and the other that quality' [230b9] means that if one thing has two things as contraries in two different ways, then there is no absurdity.

230b8 For it is possible that one of the perishings is pleasant and the other painful.

Yaḥyâ:[133]
He means that if this is the case, then they are contrary in this way and they are not contrary to generation in this way, but they are contrary [to generation] insofar as they are generation and perishing. Having said: 'The natural perishing is contrary to the non-natural one insofar as they are natural and non-natural,'[134] he follows this up by saying: 'It is possible that one of the perishings is pleasant and the other painful.'

230b10 Motions and states of rest in general have the kinds of contrariety we have decribed.

Yaḥyâ:
He says that in general, as he has mentioned, the natural motion of a thing is contrary to its non-natural motion, and similarly its natural state of rest[135] is contrary to its non-natural state of rest. For instance, the upward motion of fire is contrary to its downward motion,[136] and fire's state of rest above is contrary to its state at rest below, because the former is natural for the fire and the latter is not natural.

230b12[137] These two things [sc. upward and downward motion] in general are the most proper motions in the primary and strict sense.

Yaḥyâ:[138]
He means that the motions of the elements are the most proper motions in the primary and strict sense, i.e. the motion of fire upwards and downwards, the motion of earth and the other motions [of elements] are more primary and proper motions than the motion of the composed bodies. Thus the natural motion of fire is contrary to the motion which is non-natural for it.

The upward and downward motion of wood is not a motion in the proper and primary sense. For its downward motion is natural for it on account of the fact that the earthy parts are dominant in it and this motion primarily belongs to the earth. We should not say that its downward motion is natural for it, as we know that there is air as well as earth in it. Similarly, the natural and non-natural motion of the composed bodies do <not> primarily belong to them, because the downward motion of wood does not belong to it by nature, by its being wood, but because the earthy parts are dominant in it.

Yaḥyâ:[139]

599,5 [230b21-8] Aristotle here investigates another matter, namely the opinion that what is moving naturally moves faster as it approaches its [natural] place, because this [place] accelerates the motion, as it is the proper place [of what moves]. He calls the acceleration of this motion[140] coming to a halt or coming to rest, because it proceeds by itself towards stopping.

Then he raises a problem as follows: if every state of rest which is not eternal is generated, and its generation is this kind of coming to rest – which is an accelerating motion and if things which move against the natural course finally arrive at a state of rest, which is not eternal, then necessarily that state must be generated. If it is generated, then the motion towards it must be accelerated before it comes to rest. We know, however, that the motion of what moves against the natural course becomes slower as it approaches its state of rest. Therefore that rest cannot be generated, but it is also not eternal, and this is a contradiction.

He solves this problem by saying that it is not necessary that every rest which is not eternal has a corresponding process of coming to rest or coming to a halt; this is necessary only for natural states of rest which are not eternal, because when something moves to its natural place, then the motion becomes faster as it approaches that place, because it is the proper place. When it moves against the natural course then the moving force becomes weaker as it moves away from its proper place.

600,5 **230b26** Moreover, it is thought that, when we say of something that it is coming to a halt, this is generally either identical to its moving to its proper place or it necessarily belongs to it.

Yaḥyâ:[141]

The expression 'coming to a halt', i.e. moving faster, refers either to a motion of a thing to its proper place or, if this is not the case – namely if it does not refer to the fact that when a thing moves towards its place, it is coming to a halt, i.e. moving faster[142] – it refers to something necessarily belonging to it; i.e. coming to a halt is a part of the motion of the thing to its proper place. It follows that coming to a halt is not a generation of a state of rest when this occurs in a non-natural way and in general when something moves by compulsion.

Yaḥyâ:[143]

601,9 [230b28-231a4] He raises the following problem: Motion from a certain state is contrary to being at rest in that state; furthermore, when a motion starts from a state, then the rest in that state still exists in what moves. Thus a part of it is moving towards the goal of the motion and another part is resting in the starting point. Then it would follow that two contraries are simultaneously existing in the same thing.

> **230b32** We say about this that while there is something [of the initial state] remaining, it is in some sort of state of rest.

Yaḥyâ:[144]
This is the solution of the problem: when we say that what is moving from white is still at rest in white, we mean that it has not departed completely from the white, but that it is a mixture between white and black. This shows that it is possible that in one thing there are two contraries in a certain sense, i.e. as a mixture. It is impossible, however, that they should exist simultaneously in the same way, but 602,1
he has not said this. Therefore he says that rest is contrary to motion, not in the proper sense, but in a certain sense. The proper contrary of a motion is another motion.

> **230b32** In general, part of what is in motion is in the initial state and part is in the goal to which the motion proceeds.

Yaḥyâ:[145]
This is another proof by which he explains that two contraries may exist simultaneously in two ways, for what moves is partly in the state from where it starts and partly in the goal towards which it proceeds.

> **231a2** We have explained with regard to motion and rest how each of them is one and which are contrary to which.

Yaḥyâ:[146]
He has explained how motion is one; he has not explained this with regard to rest, but what he has said about motion is also applicable to rest.

[Aristotle 6.1]

> {231a21-b18 Indivisibles, e.g. points, cannot make up a line since they cannot be arranged continuously, contiguously or successively.[147]}

Yaḥyâ and Abû 'Alî:[148]
Since Aristotle at the beginning of this book repeats the definition of 606,6

certain things which he has defined before in the fifth book, we know that this book links up with the fifth book. His aim in the first part of the book is to explain that a magnitude is not composed of points, that time is not composed of instants ('nows') and that motion is not
10 composed of jerks.[149] For what a point is in a magnitude, an instant is in a time interval and a jerk in a motion.

He says that if a magnitude were composed of points, then for a line which is composed of ten parts it would follow that a fifth of it is a line composed of two points. Because this line[150] is continuous, the extremes of the two points must be one, for that is the definition of continuous. Thus a point would have an extreme, and an extreme of something is different from the thing itself. Thus a point would have in itself something which is an extreme, and something else which is
15 not an extreme. If that were the case a point would not be indivisible.

Then Aristotle explains this by means of division as follows: If the continuous were composed of points then it must be divisible into these, for what is composed of something must be divisible into it. If the continuous were divisible into two points, then it would be composed of them and the extremes of its parts would have to be one and the same.

Yaḥyâ and Abû 'Alî:

607,15 He gives another proof, arguing that if a magnitude were composed of points, then these points would either have to be continuous – this has been proved to be impossible – or touching, without being continuous. If they were touching[151] then they would have to touch one another as wholes. In that case they would coincide since there is no distance between them and they do not make up a magnitude at all,
608,1 for one point is at the same place as the other. Then no magnitude would arise, unless part of one point were in contact with part of another or part of one point were in contact with another as a whole, but all this is impossible as a point does not have parts.

Aristotle says[152] that the continuous must be divisible and that its parts must be different in places. He specifies this proposition by
5 adding 'in place', because things may also be different in definition, not in place, as a man who is learned and white. These attributes are not different in place, but they are different in definition.

Yaḥyâ:[153]

If each point does not have magnitude, then a collection of points cannot form a magnitude, for they do not contribute to magnitude as they do not have it.

Yaḥyâ and Abû 'Alî:

16 He explains that points cannot be in succession to one another,

because things are in succession if there is nothing of the same kind between them. Between each pair of points, however, there is a line, because they cannot be touching, as we have explained. Thus there is a line between them. It is possible to point out an infinite number of points on this line. Therefore between two points there is always something of the same kind, so they are not in succession.

Lines may be in succession, for although it is possible to indicate points between any two lines, points are not of the same kind as a line.

Yaḥyâ:[154]

Having shown that the continuous cannot be composed of points by 609,13
means of composition, he also shows that this is the case by means of division as follows: If the continuous were composed of points, it would be divisible into points; for what is composed of something is divisible into it. If a line were divisible into points it must be divisible either into divisibles or into indivisibles. That it is divisible into what is always further divisible, is what we maintain; if it were divisible into indivisibles, then these cannot be touching, and the continuous is [composed of what is] touching. Thus the continuous is not divisible into indivisible parts, so it is not composed of indivisible parts.

{231b18-232a17 Similarly, time is not composed of instants, nor motion of jerks.[155]}

Yaḥyâ and Abû 'Alî:

Having shown that magnitude is not composed of indivisible things 613,4
he now starts to make clear that motions are not composed of jerks and that time is not composed of instants – for an instant is indivisible and so is a jerk, because what a point is in a magnitude, is something else[156] in a time interval and again something else[157] in a motion. He has explained in the fourth book that for magnitude, motion along that magnitude, and time, all properties which hold for one of them, hold for the others, namely properties such as divisibility and continuity.[158] If magnitude is continuous, then the motion along it and the time which measures that motion are also continuous. If magnitude is divisible, then time and motion are also divisible.

He explains that motion is not composed of indivisible parts, because if motion were composed of indivisible parts, then it would necessarily follow that the magnitude along which the motion occurs is also composed of indivisible parts.[159] Suppose the magnitude is ABC and the motion along it is DEF; suppose the moving object is G. Let A be the part of the magnitude which is not divisible. Then this part must be traversed by a partial motion which is indivisible; let D be that part. For suppose D were not indivisible, then G would traverse the indivisible part A by the partial motion D, which is further divisible into motions. It would follow that either G traverses A [completely] with one of these motions, which are parts of the motion D – then part of a motion would traverse the same magnitude as the whole motion, which would mean that part of a thing is equal to that thing – or G traverses with that part of D a magnitude less than A – then A would have become divisible, whereas it was assumed to be indivisible. Thus, part D of the motion DEF is indivisible. The same can be shown for parts E and F, for we assume that parts A, B and C of the magnitude are all indivisible and that G traverses part A with the partial motion D, part B with the partial motion E and part C with the partial motion F, as G traverses the whole magnitude ABC with the motion DEF.

It is evident that the meaning of 'has moved'[160] is different from 'is 614,1 moving', because 'is moving' is a process and a course, whereas 'has moved' means a termination of a course. G must either be moving along part A with the partial motion D – then D is divisible, because 'is moving' implies the notions 'from' and 'towards' or 'beginning' and 'end' and these are different things and two distinct parts – or it must have moved and be moving at the same time – this is similar to the case in which Zayd is walking from Basra to Baghdad while being in the condition of having walked and having arrived – or it must have traversed A with the partial motion D without having been moving before – this is impossible in the same way as it is impossible that Zayd has walked from Basra to Baghdad without having been walking before.[161] If G has moved along A, then it is not moving, for 'has moved' has a different meaning from 'is moving'. Everything which admits of motion and is not moving, is at rest. Thus G will be resting in part A, and similarly in part B and part C, so G will not have moved while the motion has taken place, which is impossible. Then the motion would consist of states of rest; all this is absurd.

He says 'it has covered the distance' [232a2, 232a10] instead of 'it has moved'.

232a3 ... but it was in an intermediate state.

This means that what moves is partly in the state from where it starts to move and partly in the state towards which it proceeds.

Yaḥyâ [ad 232a15]:[162]
Each of the parts D, E and F of which the motion DEF is composed must either be a motion – then it is a motion in which no motion is taking place, as G has moved along each part of the magnitude ABC without having been in motion – or not a motion – then the motion

DEF would be composed of rests and would consist of rests and not of motions.

[Aristotle 6.2]

{232a23-232b20 Because of the continuity of magnitude it is possible to distinguish between motions of different speeds.[163]}

Yaḥyâ and Abû 'Alî:[164]
Having shown that magnitude is always divisible, and that the same 619,16
holds for motion, he intends to make clear that the same also holds for time. He explains this by means of the fact that time, motion and magnitude are the same in respect of continuity and infinite divisibility, because when one of these is continuous and divisible, then the other two also have these properties. As magnitude is always divisible, and motion too, time must also be always divisible.

He first brings forward three things which are evident.[165] First, 620,1
what moves more quickly covers in a longer time a distance which is greater than the distance covered in a shorter time by what moves more slowly. Second, the quicker covers in the same time a distance which is greater than the distance covered in the same time by the slower. Third, what moves more quickly covers in less time the same distance as the distance which is covered by the slower in more time. He also says that the quicker covers in less time a greater distance. This is true only if one assumes a certain velocity and a certain distance. Then it is possible that the quicker covers a greater distance in less time, but it is also possible that the distance covered is the same; so it is not necessary that all quicker motions cover a greater distance in less time and not an equal distance, whatever the distance may be.

These three things become clear from what he says in his book. Their explanation is clear, for if the quick one and the slow one move during an equal time, then it is not possible that the slower moves a greater distance, because slowness is not more effective in covering a distance than quickness. Nor does the slower cover an equal distance, because quickness is more effective in covering a distance than slowness, so it must cover a greater distance. If the quicker covers a greater distance in an equal time, then *a fortiori* in a longer time it covers a greater distance. If the time is shorter, then it may cover a greater distance or an equal distance.

If A covers a magnitude B in a time T, and G is slower, then G covers a part C [of B] in the time T; for B is divisible, as every magnitude is divisible. A, being the quicker, covers a part of B, namely D, in a part of the time T. The same may be said of all extensions which may be covered by faster- and slower-moving objects, because all magnitudes, irrespective of their extensions, are divisible.

By his phrase 'They are all divisible' [232a20] – sc. the magnitudes – he means that they are always divisible *ad infinitum*. Some people have thought that his phrase 'they are all divisible' refers to time; i.e. time is wholly divisible. This is not correct, because Aristotle says after this: 'Thus, time is divisible' [232a21].

Someone might ask: Why does Aristotle say: 'Every magnitude is
621,1 continuous, so the quicker necessarily covers a greater distance in an equal time' [232a24] and what is the connection between the two [parts of the statement]? The answer is that he says this because if some magnitude were not divisible, then it would not necessarily follow that the quicker covers that magnitude in a time which is shorter than the time in which the slower covers it, but the slower and the quicker would cover it in the same time. For it would not be possible to say that the slower covers a part of that magnitude in the same time as the time in which the quicker covers the whole magnitude.

One may clarify Aristotle's explanation that the quicker covers the same distance in less time as follows: Assume that the quicker covers a distance of one cubit in one hour, and the slower covers in that hour part of a cubit, say half a cubit H; on account of what was assumed, the quicker will cover half a cubit in half an hour.

232b5 It [sc. the quicker] has covered in less time a greater distance.

Yaḥyâ:
This is not completely true; it is true only under the circumstances he has assumed, namely that [for instance] the quicker covers in half an hour the distance CD, which is two thirds of a cubit; then the quicker covers in half an hour a greater distance than the slower, an hour being the time of motion of the slower [to traverse half a cubit].[166]

One may clarify his explanation that the quicker covers in less time the same distance as follows: The quicker covers a longer distance in less time than the slower covers that distance; and the quicker covers a longer distance in more time than a shorter one. Thus if the quicker covers a cubit in one hour, then it covers part of it, say half a cubit, in half an hour. Let us assume that the slower covers half a cubit in more than an hour. Then the quicker and the slower have both covered <half> a cubit, i.e. the same distance, while the time of the quicker, i.e. half an hour, is less than one hour, which again is less than the one hour and something [of the slower]. Thus the quicker has covered the same distance in less time.

Note that when Aristotle first mentioned that the quicker covers in less time a greater distance and now says that it covers in less time the same distance; he lets us know that he does not mean to say that all quicker motions in all circumstances cover a greater distance in less time, because if that were the case it would never cover an equal distance. Similarly, if it always covered an equal distance, it would in other circumstances never cover a greater distance.

{232b20-233a12 Time and magnitude are infinitely divisible.}

Yaḥyâ and Abû 'Alî:

Every motion occurs in a time interval, and in every time interval a 624,2
motion is possible. For every moving body it is possible that there is a quicker and a slower one. So it is possible that there is a quick motion in every time interval. The quick motion divides the time, and the slow motion divides the magnitude. Thus every time is divisible and every time is continuous, because the continuous is that which is always divisible into things which are *not* divisible.

The explanation[167] that every motion occurs in a time interval is that motion is from a certain place to another place, and between these the motion occurs. Therefore one necessarily conceives together with motion a beginning, an end and what is in between. This is how the concepts past, present and future get their meaning. They are the parts of time. Therefore every motion is in a time.

Because every time has these parts, and motion necessarily has these parts also, it is possible that in every time interval there is a motion, as every time interval has a past and a future, which are necessary attributes for every motion; they are constituent factors of the concept of motion, i.e. they are the 'from where' and the 'towards which' considered from an intermediate position between the 'from where' and the 'towards which'.

If there is a motion possible in any time interval, and if for every moving object it is possible that a quicker and a slower exists, then it is possible that in every time interval there exists a quick motion and a slow motion.

233a8 If this alternation always holds and the use of this 625,14
alternation always leads to a further division, then it is evident that every time must be continuous.

Yaḥyâ:[168]

He means that one should take the slower after the quicker and the quicker after the slower.

233a10 At the same time it becomes clear that every magnitude is also continuous, for time and magnitude are divisible by the same and equal divisions.

Yaḥyâ:
Both time and magnitude are divisible by equal divisions, because when we turn from the slower to the quicker, the time is divided, and when we turn from the quicker to the slower, the magnitude is divided. It is on this account that each of them is divided by the other into equal parts.

{233a13-31 The infinite divisibility of time and of magnitude are implied by one another.}

Yaḥyâ and Abû 'Alî:[169]
627,12 What we have discussed – namely that if either time or magnitude, whichever one it may be, is continuous and divisible, then the same holds for the other – is a fact which is also known to the general public. For when they say that this distance is covered in one day, then they say that half the distance is covered in half a day, a quarter in a
628,1 quarter of a day and one eighth in one eighth of a day, so they divide the time in accordance with the division of the distance. They also say that if the distance has been covered in one day, then in half a day half the distance will be covered and in a quarter of a day a quarter of the distance, so they divide the distance in accordance with the division of the time. It follows for time and distance that if one of them is infinite, then the other one is infinite too.

Something[170] may be said to be infinite in two ways. First, it may be infinite in the sense that it has no end or extreme. If the time in which a magnitude is covered were infinite in that sense, then the same would hold for the magnitude and if the magnitude covered in a certain time were to be infinite in that sense, then the same would hold for the time. Alternately, it may be infinite in the sense of being infinitely divisible. If the time in which a magnitude is covered is always further divisible, then the same holds for the magnitude and if the magnitude is always further divisible, then the same holds for the time too.

What holds for one of them holds for the other, because if magnitude is continuous and there is a motion along that magnitude, then this motion is also continuous, as the motion is along that continuous magnitude and traverses it; therefore, if the magnitude is continuous, the motion is also continuous. Since time is a number of motion and one of its conditions, it too is continuous. So time must be continuous if magnitude is continuous.

Similarly, if we start with the knowledge that time is continuous, we conclude that motion is continuous, because time is a number for it and in accordance with the continuity of what counts (number), what is counted will be continuous too. If motion is continuous, then the magnitude is continuous, because the motion traverses it. It has become clear that magnitude must be continuous if time is continuous.

The fact that divisibility may be infinite in these two senses provided Zeno with the sophism he used in refuting the existence of motion. As we have distinguished the infinite in this way, it is easy to understand the quibble.

Zeno has used[171] the following argument: If motion were to exist, 629,1
then the infinite would be traversed in a finite time; this is impossible, and therefore the existence of motion is impossible. He says that motion implies that the infinite is traversed, because there is an infinite number of points on the magnitude that is traversed, for a magnitude is infinitely divisible.

We say to him that the magnitude is infinite, but not in the sense that it has no limit, and the same holds for the time. Both magnitude and time are infinite in the sense that they are infinitely divisible, not in the sense that they have no limit. If you mean that it is not possible for what has no limit to be traversed in a time which has a limit, then this is true. We do not state that, but we state that if the magnitude had no limit, then the time in which it were covered would have no limit either. If you mean that what is infinite in divisibility cannot be traversed in a time which is finite in divisibility, then that is what we assert, for we maintain that time is infinite in divisibility, in the same way that magnitude is infinitely divisible. So if something traverses an infinite magnitude, it traverses that magnitude in a time which is infinite in the sense in which the magnitude was said to be infinite. As for the points which are on the magnitude, indeed there is an infinite number of them, but they do not actually exist. The traversing of a distance is something actual; therefore the traversing does not apply to the points, nor does the coming-into-contact apply to them.

233a23 [It is impossible for a thing] to come into contact with each one of an infinite number of things.

Yaḥyâ:
He means that it does not come into contact with each of the points which are on a magnitude.

{233a31-233b15 Finite distances must be traversed in finite time. An infinite distance would require an infinite time.}

Yaḥyâ and Abû ʻAlî:
Since Aristotle and Zeno maintained that it was impossible for the 631,14

infinite to be traversed in a finite time, and Zeno left this without a [correct] proof, Aristotle wanted to demonstrate this, and he also demonstrated that a finite magnitude cannot be traversed in an infinite time. He first gives the proof of the latter proposition as
632,1 follows: Suppose that the finite distance A is traversed in an infinite time B. Then take from the time B a finite part D. In a finite time a part of A is traversed, whichever part it is. Suppose in that [finite time D] a part C of the distance A is traversed. Every finite quantity has a proportion to any other finite quantity, either a proportion of one third, or one fourth, or some other proportion. So part C has a proportion to A, and it will either be an exact measure of it,[172] or not. Now let us take another time such that the proportion of D to it is equal to the proportion of C to A. Let this time be E. Then the proportion of C to A is equal to the proportion of D to E, and conversely the proportion of D to E is also equal to the proportion of C to A, i.e. one fourth or one third, or some other proportion. It follows that the complete distance A is traversed in the time E, because when one fourth of A, which is C, is traversed in the time D then one half of A must be traversed in a time which is twice D, three-quarters[173] of A in three times D and the whole distance A in four times D. The time which is four times D is E, so in E the whole distance A is traversed. As D is a finite time, and has a proportion to E, E must be a finite time, because the finite has no proportion to the infinite. This being the case, then if the distance A was traversed in an infinite time, the same distance would be traversed by equal motions, in a finite time and in an infinite time.

Someone might say that if the distance A is traversed in an infinite time, then part of A may also be covered in an infinite time. One should reply by saying that if we take a finite part from the time B, then we can traverse a part of A in that time, because we observe by our senses that in a finite time a certain distance is traversed.

233b2 This part will be either an exact measure of AB or it will be less or greater; it makes no difference.

Yaḥyâ:[174]

He says that the magnitude of which we have said that it is a part of AB and which is traversed in a part of the infinite time, must be either an exact measure of AB – so that if we multiply it the result exhausts AB and is neither more nor less; for instance, if that magnitude is one cubit and the whole distance AB is seven cubits, then after having multiplied the cubit by seven the result is neither less nor more than AB, but exactly equal to it – or there remains a remainder, which is smaller or greater. Suppose for example is that this magnitude is two
633,1 cubits and AB is seven cubits. If we multiply the two cubits by three,

one cubit remains [as a remainder], which is smaller than two cubits, and if we multiply the two cubits by two, three cubits remain, which is more than two cubits.[175] In each of these cases this magnitude necessarily has some proportion to the seven cubits.

> **233b4** If the magnitude is equal, i.e. equal to BE, it will always be traversed in an equal time.

Yaḥyâ:
He means that if the magnitude BE is traversed in the time CD and the whole magnitude AB is finite, so that the part BE measures it out, either by three times or by four times, then the following parts are traversed in equal times. For what comes after BE are parts which are all equal to BE, and if parts are equal and are traversed by equal motions, they are traversed in equal times. It follows that the whole magnitude is traversed in four times the time in which the magnitude BE is traversed, and four times that time is finite; that was our supposition.

> **233b6** For it is divisible into parts which are equal to the parts into which the magnitude is divisible.

Yaḥyâ:
This means that if the magnitude is divided into equal parts, the time is also divided into equal parts. Also, if the magnitude is finite, the time in which that magnitude is traversed is also finite.

> **233b7** Furthermore, if not every magnitude is traversed in an infinite time ...

He says:[176]
One cannot object that a part of the finite magnitude which is traversed in an infinite time must always be traversed in an infinite time too, because one cannot say that every magnitude is always traversed in an infinite time. A finite magnitude may be traversed in a finite time; this is evident and may be observed by our senses.

> **233b11** It is obvious that the part BE is not traversed in an infinite time, if time is taken to be limited in one direction; for as the part will be traversed in less time [than the whole] this time must be finite, as it is limited in the other direction.

He says:[177]
The part BE of AB must be traversed in a finite time, even if it was 634,1
assumed that the time in which the magnitude AB is traversed is

infinite, because this time is necessarily finite in the direction of its beginning, since the traversing of the magnitude AB has started at that instant. If that time is finite at its beginning, it must be finite at its end, i.e. the time in which the magnitude BE is traversed. For if that time were infinite, and the time in which AB is traversed were infinite too, while at the same time longer, then there would exist something greater than infinite, which is impossible. It is not impossible to suppose that the time is infinite and then to take a finite time from it, but the impossibility is that two times should be infinite while one is longer than the other.

Yaḥyâ:[178]

635,3 He intends to say that for time, motion and magnitude, any one of them is divisible on account of the divisibility of another. If magnitude is continuous, and the continuous is divisible into divisible parts, and motion is continuous on account of the continuity of magnitude, then motion must be divisible too. Similarly, time also must be divisible, because it is necessarily continuous on account of the continuity of motion.

Or he means by this that in the same way as magnitude cannot be composed of points, motion cannot be composed of indivisible parts, because motion occurs along a magnitude. Furthermore, because motion occurs in time and may be quick or slow, it is also necessary that time is not composed of instants.

Yaḥyâ and Abû 'Alî:

636,15 This [233b15-32] is another proof that nothing which is continuous can be divided into indivisible parts; it runs as follows: If continuous things were divisible into indivisible things, then it would follow that the indivisible would be divisible.[179] For in every time a quicker and a slower motion is possible and the quick motion covers in the same
637,1 time a greater magnitude than the slower motion. Let us suppose a time which is composed of three parts which are all indivisible and let these parts be A, B and C. Let the moving object which is quicker cover a magnitude composed of three parts K, L and M in this time. Let us further suppose that the ratio of the quick motion to the slow motion is one and a half, on account of the fact that the slower motion in the time ABC covers a line composed of two parts. Then it follows that it covers one of these two parts in one and a half time-parts. So the time-part is divisible, whereas at the same time it was assumed to be indivisible.

If someone were to say: 'Why do you not assume that the quicker motion covers the three parts in six time-parts?', then we would say to him: 'If we assume this, then it follows that in each of these time-parts half of a magnitude-part is covered, whereas at the same time we have supposed that the parts of the magnitude were indivisible.'

[Aristotle 6.3]

{233b33-234a24 Instants are indivisible.}

Yaḥyâ and Abû 'Alî:[180]

He assumes that the now which has no extension is indivisible, 640,13 because this is evident. For if it were divisible it would be a [stretch of] time, not a limit of time. If it were divisible it would have an extreme and a limit, and a limit is different from what has a limit. Also, what can be said about the limit of the now is the same as what can be said about the now itself, namely that it must be divisible like the now and have a limit, and this limit will again have a limit and so on *ad infinitum*. All these limits would be times, and there would be an infinite number of them. A consequence would be that there would be an infinite time between day and night.

Furthermore, time has one dimension, like a line. As the limit of a line is indivisible, the same must hold for time. A surface has two 641,1 dimensions, so its limit is necessarily divisible; this limit is a line, which is divisible.

Then Aristotle explains that the now, which is the end of past time and the beginning of future time, is one and the same thing. For if the now were two things, one of them being the beginning of the future and the other the end of the past, then they must be either continuous or touching or separate. When each of the nows is indivisible, they cannot be continuous, because the continuous cannot be composed of indivisibles. They cannot be touching either, because things are touching when their extremes are together, and what has an extreme is divisible. Thus things which are indivisible cannot be touching. They cannot be separate either, because time is continuous, so there must be a time between these two nows, otherwise time would not be continuous. If there were a time between these nows and if each time is divisible, then this time must be divisible, whereas at the same time it is the beginning of the future and the end of the past. Then it would follow that the same time would simultaneously be past and future. It would be past, because it is the end of past time, and it would be future because it is the beginning of future time.

Because this 'extended' now, i.e. this now, is divisible into a past time and a future time, the part which is past time must be past and future: past, because it is the end of past time and future because it is the beginning of future time. The part which is future time will also be past and future.

Furthermore, if the point of separation between past time and

future time were two nows, one of them being the end of the past and the other the beginning of the future, whereas at the same time there must be something between these nows which is described by the same name (*sunônumon – muwâṭi'*) as these nows, i.e. something similar and like – and that is a stretch of time – then this time between these nows is not past, nor future, for only what is separated
642,1 by these two nows is past time and future time. If this time between these nows is neither past nor future, then it cannot be time, whereas at the same time it was supposed to be time; this is impossible. Thus, if the now were two, then only these alternatives exist and they are all impossible. It has been refuted that there are two nows, so the now must be one. These absurdities also follow when one supposes the now to be divisible, therefore it cannot be divisible.

Aristotle says that the now is the beginning of the future and the end of the past and he investigates, taking into account this proposition, whether it is one and the same thing [or two. If it were two nows, there must be a time between them, and] if there is a time between them, it is divisible, because time is divisible. Then something of the past, which has been, is in the future and something of what is in the future has been [in the past]. For these two nows, between which there is this time, are, together with that time, the beginning of the future, for both are the beginning of the future; from this point of view that time is future. Because they are, together with the time between them, the end of the past for both nows are the end of the past that time must be past.

Furthermore, if the now is two, and if there is a time between them, and every time is divisible, then this now cannot be a now by itself, in the proper sense, but it will be a now which has an extension, and we have assumed that we would discuss the now which has no extension.

233b34 In all time there must be something of this description.

Yaḥyâ:
He means that whatever now one takes, it is a separation between two days, or a separation between two hours, or between two months. It is of this description; i.e. it has no dimension and it is not divisible.

Aristotle's saying [234a3]: 'If it is shown about this … ' means: if it is shown about the now that it is one.

234a13 For where the point of division is, there past time is marked off from future time.

Yaḥyâ:[181]

Because between the end of past time and the beginning of future time there is a now, which has an extension, not an indivisible now … .[182]

> **234a16** Furthermore, a part of the now will be past and a part will be future, and it will not always be the same part that is past or future, and the now will not be one and the same, for the time may be divided at various points.

He says[183] that if this now has an extension, then a part will be past and over, and a part will be future. It does not mean that the now has been over or will be in the future, but that a part has been over and a part will be in the future. This now is not one and the same, because it is divisible into past and future, and past and future are not the same things.

His statement [233b33]: 'The now also is necessarily indivisible, that is, not the now which is called 'now' in virtue of something else, but the now in its proper and primary sense' means[184] that the now about which he is speaking is the now which has no extension. For the now which has extension is called 'now' in virtue of something else, namely because of its nearness to the now which is indivisible; and because it is accidental for it that it is near to the indivisible now, it is not a now in the proper sense, but by accident, so it is not a now in the primary proper sense.

His statement [233b35]: 'For it (the now) is a limit of the past, no part of it belonging to the future, and also of the future, no part of it belonging to the past' means that the now is indivisible. Therefore one cannnot say that part of it is past and part of it is future, but it is one thing, which belongs to the past as well as to the future.[185]

His statement [234a9]: 'It is necessary that between the limits there is something described by the same name (*sunônumon – muwâṭi'*)' means that between two nows which are separate there are other nows which are described by the same name as these two nows. This must be the case, because there is a stretch of time between them, and in every stretch of time there is potentially an infinite number of nows. There must be a stretch of time between them, because time is continuous, so it is impossible that there is no time between two separate nows.

His statement [234a13]: 'For where the point of division is, there past time is marked off from future time' means[186] that this now, which has an extension, and which is a stretch of time – and time is divisible, being partly past and partly future – is partly past and partly future.

{234a24-234b9 Nothing moves or is at rest in an instant.}

Yaḥyâ and Abû 'Alî:

646,6 If something moved in a now, it would necessarily follow that the now is divisible, because every motion may be faster or slower. Suppose that the faster motion covers a distance A in that now; then the slower will in the same now cover a distance less than A, call it B. Because the faster covers the distance before[187] the slower has covered it in less time than the slower, the faster must cover the distance B in less than this now. Thus the now is divisible, which is in contradiction to what we have shown.

If there were motion in a now, then the same thing, being in the same condition, would be moving and resting, because the same now is the end of the preceding hour and the beginning of the next hour. It is possible that the thing is moving for the whole of the first hour – and when it is moving in the whole of that hour, it will also be moving in the now which is the end of that hour – and that it is at rest for the whole of the second hour – then it will also be at rest in the now which is the beginning of that hour, which is the now about which we said it is moving in it. So it must be moving and be at rest at the same time.

If the now does not admit of motion occurring in it, it is equally impossible that something could rest in it. For what is at rest is what has stopped moving, while it is disposed to move at the time at which it is disposed to move, at the place at which it is disposed to move and in the manner in which it is disposed to move. As the now does not admit of motion occurring in it, there can be no rest in it.

647,1 Our expression 'disposed to move' is [used] as [it is used in the claim that] a stone is not disposed to walk, so that one cannot say that it is at rest from walking. 'The time at which it is disposed to move' is [used] in the sense that we do not say of a young child that he is resting from walking,[188] because he is not disposed to do that at that time. By 'the place in which it is disposed to move' we refer to things like air and earth, for we do not say that fish are at rest from flying and moving in the air, because fish are not disposed to move in this place. By the expression 'in the manner in which it is disposed to move' we mean the manner of the motion. For instance, we do not say about a man that he is at rest from walking when he is not walking on his hands, for he is not disposed to walk on his hands. We say that he is at rest from walking when he is not walking on his feet. We do not say of what is disposed to change in another way than walking that it is resting because it is not walking. Nor do we say of something which is not disposed to have a circular motion that it is resting[189] when it is not moving in a straight line. Furthermore, what is at rest is what is in the same condition at a certain now and before that. But the now does not contain a 'before', thus there cannot be a resting in it.

[Aristotle 6.4]

{234b10-20 Everything which moves must be divisible.}

Yaḥyâ and Abû 'Alî:
What changes is either wholly and completely at the starting point, 649,2
or it is completely at the goal, or it is both at the starting point and at the goal, or it is neither at the starting point nor at the goal, or a part of it is at the starting point and another part is at the goal. If it is at the starting point, then it is not changing, because it has not started to change, and what is changing has begun to change. If it is as a whole at the goal, then it has changed and it is not changing. If it is at the goal as well as at the starting point, then it would simultaneously be white and black, when we assume that it changes from one of these states to the other. If it is not at the starting point, nor at the goal, it is not[190] correct to say that it is changing from white to black, because if it is in neither of these states the transition is not from one of these states to the other, as its changing from one state to the other means that it gradually turns from one of these states to the other. What remains is that part of what is changing is [at the starting point and part of it is] at the goal; from this it follows that what is changing must have parts.

We do not mean, by our saying that a part of it must be at the goal and a part at the starting point, that a part will be black and a part white. We mean <by starting point and goal> the intermediate states, such as grey and green and other intermediate states between white and black, of which there are an infinite number. If the motion from white to black is continuous, then each of those intermediate states is a goal in relation to any state before it, and a starting point in relation to any state after it. If the motion stops at grey, or any other intermediate state, then the state at which the motion stops is the goal, because nothing comes after it. If the motion continues again towards black, then the motion is from grey, which is a starting point as far as there is some white in the grey, not as far as there is some black in it. For if the moving object were to move from grey as far there is some black in it, then it would change and move from black. If that were the case, then the transition would not be from black to purely black, but it would necessarily go towards white.

Alexander has raised a problem as follows: Is it not true that a

25 change may occur instantaneously, not part by part, as milk coagulates instantaneously, as Aristotle has said,[191] and a face which becomes pale instantaneously when it is exposed to the sun? Then it is not the case that one part becomes pale and then another part, nor that one part of the milk coagulates and then another.

650,1 He has solved the problem, saying that Aristotle here only discusses what does not change instantaneously, such as the burning of wood and such kinds of process. What changes instantaneously is also divisible, for such a thing is nothing but what changes in all its parts together, not one part after the other.[192]

Yaḥyâ:

9 The divisibility of changing things may be explained in another way, as follows: What changes is potentially at the goal. When it has arrived there, it has come to be actually at the goal and potentially at the starting point. When it is potentially at the goal, then it is the matter which is potentially there. What consists of matter is divisible, because matter is the cause of the divisibility of the forms; the forms themselves are one, but they become divisible on account of the matter. Then they are divided and become many, and therefore the form becomes many.

15 If someone should ask how it happens that the rational soul, which by itself is actually knowing, becomes potentially [knowing], then the answer is that the soul gets potentiality when it comes to be connected to a body. It obtains potentiality on account of the matter, as one sees, for when it comes to be connected to a body, it becomes subject to something like forgetfulness, so that the knowledge becomes potential, and the soul sometimes has knowledge, sometimes not. When the soul is separated from the body, then it is actually knowing,
20 according to Plato. Thus potentiality occurs to it on account of the matter.

Yaḥyâ and Abû 'Alî:

654,9 [234b21-235a13] Having explained that everything that is in motion[193] is divisible, Aristotle explains here that every motion is divisible in two ways, first, according to the division of the moving body, and second, according to the division of the time in which the motion occurs.

He explains the divisibility according to the parts of the body as follows: Suppose the body is ABC and the motion is DF. Because ABC is moving, its parts are necessarily moving, because it is impossible for the whole to be moving if its parts are not moving. Let us divide
15 ABC into two parts, denoted by AB and BC; let AB move with a part DE of the motion DF and BC with the part EF, then the two parts AB and BC move with the parts DE and EF respectively, while the

whole motion DF is the motion of the whole body ABC. Thus the motion has been divided according to the parts of the body.

Suppose[194] someone says that the motion DF is not the motion of the body ABC. If one were to say that, then either the motion DF would be the motion of no moving object, which is impossible, or it would be the motion of one of the parts AB and BC, either the one or the other, or DF would be the motion of something else, i.e. something which is not the body ABC. It cannot be the motion of the part AB, because we have assumed that the motion of AB is the part DE, and 655,1
it is impossible for one thing to have two motions and it is impossible for a motion to be a motion for something of which a part moves with the same motion. If the motion DF were a motion for something other than ABC, then its parts would be motions for parts of that other thing, because if something has a motion, then its parts have motions which are parts of that motion. Then it is impossible for the parts of the motion DF to be motions for the parts of the body ABC as well, for one thing cannot have two motions. But we have assumed that the parts of the motion DF are motions for the parts of ABC. Thus DF is the motion for the body ABC and its parts are motions for the parts of the body and the motion has been divided according to the division of the body.

Furthermore, if the whole motion DF were not the motion for the body ABC, then this body must have another motion, say GH. Because the whole body moves with this motion, its parts must be moving too, because it is impossible for a whole to be moving without its parts being in motion. Let us posit two parts of the motion GH, one of them being the motion GI and the other the motion IK, such that the parts of the body, i.e. AB and BC, move with these motions; for if a thing has a motion, the motion of its parts are parts of that motion. Then either a part of the motion [GH] will surpass the sum of the parts [GI and IK], or the parts GI and IH will fall short of these parts, or they will neither surpass them nor fall short of them.[195] If [the sum of] both parts [GI and IK] is surpassed by a part of the motion [GH], then the surplus must be either a motion of nothing – which is impossible – or a motion of the whole body ABC – which would imply that one thing would have two motions and that part of a motion would be a motion for a body which also has the whole motion as a motion – or a motion for one of the parts of the body AB or BC – but each of these parts has another motion, so that there would be two motions for one thing – or a motion for something else than the body ABC – then there would be one continuous motion for two discontinuous things; for the whole motion GH is continuous and the body ABC together with that other body is discontinuous.

If there is a remainder from the body ABC after having taken the motions GI and IH, then this implies that the remainder would be moving and be at rest at the same time: it would be moving because we have supposed that the motion GH is the motion of the whole body 656,1 ABC and it would be at rest because no part of the motion GH is left that has not been exhausted by the parts of the body coming before the remainder.

Thus the [motion of the] parts AB and BC do not fall short of the parts of the motion GH, nor do they surpass them. Consequently these parts [of GH] are the same as DE and EF, otherwise one thing would have two motions. If the parts of the motion DE are the same as the parts of the motion GH, then the motion GH is the same as DE, because its parts are the same as the parts of the motion GH.

Yaḥyâ:
Someone may raise a problem,[196] saying that if motion were divisible according to a division of the moving body, this would imply that if a grain moved a distance of one hundred cubits and we divided this distance in our mind into ten parts, then with each part of the motion a tenth of the distance would be traversed; this would imply that each part of the grain moves with its own part of the motion, so that a certain part of the distance would be traversed only by a certain part of the grain and another part of the distance by another part of the grain. It is well-known that each part of the grain traverses the whole distance with the whole motion and that one cannot say that each part traverses a part of the distance. The proposition that the motion is divisible according to the parts of the moving body would imply that each part of the moving body only traverses its own part of the distance.

Similarly, when a body consisting of five parts becomes white, this would imply that each part gets a portion of whiteness, so that the form of whiteness which the body acquires is completed by adding up the portions of whiteness which are in the parts of the body. But this is not the case, because each part of the body gets the same form of whiteness as the whole, and it is not the case that[197] a part is less white [than the whole], so that[198] the whiteness is completed by adding the whitenesses of each part; the form of the whiteness is one [for all parts].

The one who raises the problem says that what Aristotle has said is correct for growth and diminution. For if someone grows one cubit, then his leg grows an inch, and similarly his spinal column, his belly and all the other parts of his body, in such a way that these partial growths add up to one cubit. The same holds for diminution.

The solution of the problem is that, indeed, each part of the grain 657,1 traverses the whole distance, for when the whole grain traverses the whole distance, then its parts must also traverse that distance. But for each part of the grain, the traversing of that distance is particular to it and it cannot be the same as the traversing of the distance by the whole. Thus, although a part traverses the whole distance, the traversing of the distance by a part is not the same as the traversing by the whole, but it is a part of it. The motion is not subject to division by itself, but accidentally, i.e. in virtue of the body.

The same may be said of the parts of what becomes white. For each part gets a whiteness which cannot be the same as the whiteness of the whole. In this way the whiteness of the whole is divided into the whitenesses of the parts, not as a division of the whole into anhomeomerous parts, but into homeomerous[199] parts. Therefore the form of whiteness which is in a part is the same as the form of whiteness in the whole.

Some people have thought that each part of a changing thing has a share in the changing of the whole and that the change of the whole and the form of the whole are accomplished by the cooperation of the changes of each part. They say that it is like the haulers of a ship: each one has his share in pulling and moving and the sum of their efforts results in one motion of the ship. In the same way the whole changing body has one change, which is composed of and constituted by the changes of the parts.

Yaḥyâ:

This is not like what we are considering here. What we are considering is a division of the motion according to a division of the moving body, whereas they mention an example of a division of the motion according to a division of the movers. Also, it is not true that each of the haulers is pulling a part of the ship, as they think, but together they are pulling the whole ship. Furthermore, they cannot apply what they say to what changes instantaneously, for in that case a part gets the form of the whole and the form of the whole is not gradually realised out of the cooperation of the changes of the parts or their powers.

They want to solve the problem, saying that if a stone of one *raṭl*
covers a certain distance in one hour, then its motion in that hour is 658,1
the result of the motions of its parts, which are twelve *wiqiyya*.[200] If one assumes that each of these *wiqiyyas* does not cover that distance in that hour, whereas when they are joined together the whole covers the whole distance in one hour, then we know that this is the result of the joining together of the motions of the parts of the *raṭl*.

Yaḥyâ:

This is not convincing, because they divide the motion according to the efficient cause, i.e. the weight, not according to the parts of the body, for the *raṭl* traverses this distance in one hour in virtue of the fact that it is the weights [of the parts], and nothing else, that are joined. One may say something similar about alteration. If one cubit and two cubits change under the influence of the sun, and the change for the two cubits is quicker, then this does not necessarily occur because there are more parts in the two cubits, but because the efficient cause of the change is more powerful in the two cubits. For the sunrays are larger in number and stronger on the two cubits; therefore the change of the two cubits is quicker.

Yaḥyâ:
If one *wiqiyya* moves a certain distance in a certain stretch of time, and the *wiqiyyas* are joined together, then the motion (of the whole) will necessarily be slower, because the parts increase in number while the efficient cause remains the same; then the effect will be smaller and slower than when the cause remains the same and the number of parts is smaller. If our objector does not take into account[201] the increasing of the weight we mentioned, but restricts himself to joining the parts together, then it follows that the motion of the *raṭl* must be slower. If a certain number of parts are joined together, and the efficient cause is increased the same number of times, then the velocity must remain the same. For instance, if two *raṭl* of milk are joined together and one puts a quantity of rennet in it, then the milk is curdled in the same time as when the two *raṭl* were separate and half of that quantity of rennet was put into each of them. If one *wiqiyya* of the stone in two hours moves one cubit and then twelve *wiqiyyas* are joined together, then in virtue of the preceding argument the whole will also move that cubit in two hours.

Alexander has solved the problem as follows:[202] He says that we may divide motion according to a division of the moving body insofar as it is moving. If we assume that something becomes white, then first it acquires a part of the form of whiteness, and then it gets more and more of it until the form of whiteness has become complete. Thus the motion has been divided according to these parts.

Yaḥyâ:
659,1 This is not correct, because this is a division of the motion according to the parts of time, and that was not what we intended to discuss. Our intention is to discuss the division of motion according to the parts of the moving body.

234b29 Furthermore, if every motion is a motion of something ...

Yaḥyâ:[203]
One may think that with the word 'furthermore' (*eti de*) he begins

another proof, but this is not the case; it is a corroboration of what precedes.

235a2 If the whole motion GH may be divided into the motions of both parts, then the motion GH must be equal to the motion DF.

Yaḥyâ:[204]
The word 'equal' stands for 'the same'.

235a7 A similar result follows if the division shows a surplus.

Yaḥyâ:
He means that if there is surplus of the body when the division has been performed, then the same absurdity follows as when there is a shortage of the body at the division and there is a remainder of the motion. He does not mean by the expression 'a similar result follows' that the absurdities are completely alike, but that the argument in order to derive an absurdity runs along the same lines and that this alternative implies something similar to the preceding absurdity, namely that there is a motion which does not have a moving body. For if there remains something of the body, then this remainder must have some motion, say M. But the motion GH exhausts the whole body from which this remainder is left over; therefore this remainder does not need this motion M, so the motion M does not have a moving body.

235a10 This necessarily applies to everything that is divisible.

Yaḥyâ:
He says 'divisible' instead of 'movable', as this proposition necessarily holds for everything which is movable.

235a10 The other division is according to time.

Yaḥyâ:
Every motion is in time. It is possible for every time to find a shorter one, and the same thing will perform less motion in a shorter time. Therefore motion is divisible according to time.

Yaḥyâ:
[235a13-b5] He wants to explain that the moving body, the thing 661,10
along which the motion occurs, the time, the motion and the being-in-motion are all similar in the sense that if one of them is divided,

the others are correspondingly divided, and if one of them is infinitely divisible, the others are infinitely divisible too.

Suppose that A is the distance along which a motion occurs, B is the motion and C the time. If a moving body has this motion B, and we consider one half of this motion, then half the distance A is covered in half the time C with this half of the motion. If the body moves during half the time C, then it moves with half the motion B half the distance A. If it moves half the distance A, then it moves with half the motion B during half the time C. Similarly, when we assume that the thing in which the motion occurs is a form, for instance the form of whiteness, then if the moving body reaches the form of whiteness with motion B in the time C, and the form of whiteness is A, then with half of this motion it will reach one half of the form A in half the time C. If it reaches half the form A, then it moves with half the motion B during half the time C. If it moves during half the time C,
662,1 it moves with half the motion B and reaches half the form A.

The argument is the same for the being-in-motion. If the body moves with the motion B, then half of the being-in-motion occurs with half the motion B in half the time. If one has half the motion B, then one has half the being-in-motion in half the time. If we have half the time, then we have half the being-in-motion and half the motion B.

The division of the motion according to the division of the moving body has just been explained above. In the same way as we have mentioned here, one may discover that the being-in-motion is divisible according to a division of the moving thing, because each part of the moving body has its own being-in-motion.

Yaḥyâ:[205]
The expressions 'motion' (*kinêsis* – *ḥaraka*) and 'being-in-motion' (*kineisthai* – *taḥarruk*) either have the same meaning, or motion is like something which is like a disposition (*hexis* – *malaka*), whereas being-in-motion is the actual process.[206]

235a13 [Everything which moves,] moves along a certain thing.

Yaḥyâ:[207]
This is the kind and the form of that along which the motion occurs.

235a17 Although the divisibility of that along which the motion occurs is not of the same kind in all cases.

Yaḥyâ:
He says that not all forms along which motion occurs are divisible in the same way, because that along which local motion and growth and

diminution occur is divisible by itself, but that along which alteration occurs is divisible by accident, not by itself.

Yaḥyâ and Abû ʻAlî:
Let us suppose a being-in-motion FGH. We know that every moving body has a motion and that the parts of what moves correspond to parts of the motion. Let part FG of the being-in-motion correspond to part CD of the motion, and part GH of the being-in-motion to part DE 664,1
of the motion, then I say that the whole motion CE corresponds to the whole being-in-motion FH, because it has been established that the whole being-in-motion FH corresponds to something and that the parts of the motion CE correspond to the parts of the being-in-motion FH. If the whole being-in-motion did not correspond to the whole motion, then the parts of the motion CE must be motions of parts of that which has the being-in-motion, namely the being-in-motion FH, and of parts of FH. Perhaps one should read: and of the parts of some other moving body.[208] Then the parts of motion CE are motions of parts of something which has the whole being-in-motion FH and of the parts of some other moving body. For if the motion as a whole does not correspond to this being-in-motion, then that which does correspond to this being-in-motion must be something else, the parts of which will correspond to the parts of the being-in-motion.

Yaḥyâ:[209]
Someone might say: 'If you say that motion here is something like a disposition, then how is it possible to say that it is divisible according to a division of the being-in-motion and according to a division of time? For a disposition is something steady which does not change, irrespective of whether the moving body is quick or slow or even whether it moves or does not move, because a disposition is something which remains as it is.'

Yaḥyâ:
It is as I said, namely that motion is divisible, because it is something existing in a divisible subject, so it is divisible by accident, not by itself.

235a33 For if one takes the being-in-motion corresponding to each of the two motions, then the whole will be continuous.

Yaḥyâ:[210]
There is no difference between the whole and the parts, except that if the whole is taken as a whole, it is taken as something continuous and one. We take the parts as something divided in our minds, and therefore we conceive them as parts. Just as motion is continuous,

but is constituted by parts in our mind, also being-in-motion, even
665,1 when it is constituted by parts and then taken as one thing, may be said to be continuous. Just as the parts of the motion correspond to the parts of the being-in-motion, the whole motion corresponds to the whole being-in-motion.

> **235b1** It follows in particular that all these are divisible and that they are infinitely divisible on account of the changing body.

Yaḥyâ:
He says that as the changing body is divisible, that which belongs to the changing body, i.e. time, motion and being-in-motion, must be divisible also.

> **235b4** That divisibility primarily belongs to it (i.e. to the changing body) we have shown above. That infinity does so we shall show after this.

Yaḥyâ:[211]
He says that as infinity is predicated of divisibility, in the sense that a division may be continued infinitely, and of what has no extremes, and as we have explained above that magnitude is infinitely divisible and that the same holds for motion and time, we shall explain that if a magnitude had an infinite extension, then the same would hold for time and motion.

[Aristotle 6.5]

> {235b6-19 At the instant at which a change terminates the body is in the state towards which it was changing. Case 1: change between contradictories.[212]}

Yaḥyâ and Abû 'Alî:[213]
667,8 He explains that what has changed is at the first instant it has changed in the state towards which it was changing. He imposes the condition 'at the first instant at which it has changed' in order that one does not conclude that what has changed into black and then goes back and changes into white is still black while it is changing into white. Therefore he says 'at the first instant at which it has changed', because if what has changed to black goes back to white, it is during its return to white not in the state it had first changed to.

He shows this in the following way: As everything which changes changes from something towards something else, and changing and leaving [a state] mean the same thing – or one is a necessary

consequence of the other – it follows that if what leaves something 15
arrives at that where the leaving comes to an end, the same must hold for the change too, because what changes moves away from where it starts and from what it leaves.

Yaḥyâ:
By 'leaving' I understand nothing but 'changing' and by 'changing' nothing but 'leaving', for one implies the other.

Aristotle explains the question under discussion by means of [the case of] motion between contradictories after having stated that what is true for one kind of change, is true for all kinds of change. One knows that 20
what has changed from not-being to being must at the moment we say it has changed be either in that from where it starts or in that towards which it proceeds, or between these states. It is impossible that it is in 668,1
that from where it starts, because it has changed from that state. It is also impossible that it is in between, because there is no intermediate state between being and not-being. The only alternative is that it must be in the state to which it was changing.

Yaḥyâ:
There is no change from absolute not-being to absolute being, but change is always from what does not exist in a certain way to what exists in a certain way. For instance, the change from not-man to man 5
is a change from sperm to man and there are many intermediate states the changing thing passes through.

Thus the changing thing must pass through intermediate states and paths between the starting point and the goal, and if there are no paths and intermediate states, there is no change either, as he has informed us before. If this were not the case, then 'has changed' and 'is changing' would be the same thing. Thus 'has changed' must be preceded by 'is changing'. 10

235b13 For there is the same relation between the two in each 14
case.

Yaḥyâ:[214]
He says that 'has changed' means that the change is finished; for example, 'has left' means that the leaving has come to an end.

{235b19-32 Case 2: change between contraries.}

Yaḥyâ and Abû 'Alî:[215]
If what has changed from A has left it, then it must be either in B, 669,15
which is that towards which it is changing from A, or in something else. It cannot be in something else, say C, because C is not contiguous

to B nor touching B, but C must be between A and B. The interval [AB] is infinitely divisible, because it is something continuous, and
670,1 also the change in this interval is infinitely divisible, because change is continuous. This would mean that the changing body has changed into B, whereas it is also changing towards B. This is a contradiction. Therefore what has changed into B is in B.

Yaḥyâ:[216]
If C were touching B and the interval CB were not further divisible, then it would be possible for the changing object to be in C and to have changed into B, even if it had nothing of B yet; but it has been explained that the continuous cannot be composed of indivisible things.

> **235b20** ... for it is necessary that what has changed must be somewhere or in something.

Yaḥyâ:[217]
Having discussed change from one place to another, i.e. change in the 'where', he now starts to speak about all kinds of changes, by adding 'or in something'. By this he means change in form.

> **235b22** If what has changed into B is in something else, say C, ...

Yaḥyâ:[218]
If it is said that what has changed is in something different from the starting point and the goal, one may understand this in two ways. First, it may be in something which is between them, as when what changes from black to white is in the state of being grey. Or one may understand that it is in another genus, as when we say that what changes from black to white has become hot before having arrived at being white.

It has been explained that the first alternative is impossible. The second alternative has its own kind of refutation, namely that forms of different kinds cannot be continuous, so a change occurring in these forms cannot be continuous either. Therefore a continuous change in discontinuous forms is not possible.

> {235b32-236a7 The first instant at which a change terminates is indivisible.}

Yaḥyâ and Abû 'Alî:
672,2 Suppose the 'primary now' at which the changing object has completed its change were divisible. Let us call this now AC and let it be divided at B. Then either the changing object must have completed

its change in both parts [AB and BC] together, or it has not completed its change in one part nor in the other, or it has completed its change in one of them. If it had completed its change in both of them, while one part precedes the other, there would be something which precedes the primary instant [of the completion of the change], or the preceding and the next parts would coincide. If it has not completed its change in one part nor in the other, then it has not completed its change in the whole AC either, because the whole is nothing but the two parts AB and BC. If it has completed its change in one of them, it follows that the change has not been completed in the whole as the primary instant [of the completion of the change] and there must be something else which is prior to the primary, i.e. AB.

This [last] alternative is mentioned by Aristotle. He does not mention the possibility that it has completed the change in both parts together, because this case is evident. If all these alternatives are impossible, it follows that it is impossible for the now in which something has completed its change to be divisible, and it has been demonstrated that it is indivisible.

> **235b33** I mean by the expression 'primary' what has this characteristic not in virtue of something else.

Yaḥyâ:[219]
We say that a thing has completed its change at this hour or on this day, not because it has completed its change during that whole day or that whole hour, but because it has completed its change at the instant which is the beginning of that hour. A change is completed at an instant not in virtue of the fact that it is completed in something else to which that instant belongs, as we say that the completion of a thing's change[220] at that hour occurs in virtue of the fact that the change has been completed at the instant which is the beginning of that hour.

> **235b37** If it has been changing (*kâna yataġayyaru*) in each of these intervals – for it must either have changed, or be changing in each of them …

Yaḥyâ:[221]
By 'has been changing' (*yataġayyaru*) we must understand 'has 673,4
changed' (*taġayyara*).[222] What he intends to say is that the changing object in AB and BC either has completed its change in each of them, or is changing in each of them. But if it is changing in both of them, it is changing in the whole consisting of these parts, and if it is changing there, it has not completed its change there, whereas it was assumed that the changing thing had completed its change there.

236a2 The same argument will apply if something is changing in one part and has completed its change in the other.

Yaḥyâ:
He does not mean that in this case the same absurdity follows, but
15 that in this case some absurdity also follows.

20 **236a5** It is evident that what has perished and what has come to be must have done so in an indivisible instant.

Yaḥyâ:[223]
674,1 He means that the same argument shows that the primary instant at which something comes to be or perishes is indivisible.

{236a7-27 There is no first instant or first period of the process of changing.}

Yaḥyâ and Abû 'Alî:
675,13 He shows that there does not exist something of which one can say that the changing object starts to change in it and which has the property that the changing object has changed from rest in it. It is known that the completion of a change occurs at an indivisible instant. Now he says that if there existed such a thing for the changing object as we just have described, it must be either divisible or indivisible and both alternatives are impossible. Therefore there does not exist such a thing for the changing object.

It is not possible that it is indivisible, for if this first instant were indivisible, then it must either be the [last] instant at which the object is at rest, and then there would be an instant at which it is moving and resting simultaneously, or it is something else. In the latter case,
20 either it must be continuous with the [last] instant at which the object is at rest before the start of the motion, or it is not continuous with
676,1 it. If it is continuous with it then one indivisible is continuous with another indivisible, which is impossible. If it is not continuous with the earlier instant [sc. the last instant of the preceding rest], there must be a stretch of time between them, in which the object is neither at rest, nor moving. For it was assumed that it was resting at the former instant, and not at rest after that[224] and that it is moving only
5 from the second instant, so that between these instants it will be neither resting nor moving. Thus the claim has been refuted, that the moving object has a first instant at which it is moving, which is indivisible.

If that first instant were divisible, then it must either be changing in both of its parts, or in neither of them, or in one of them. If it is changing in both parts, then it follows that before the first instant there is something else, because it was supposed that it was changing in the whole first instant and this whole has two parts, one of them being before the other, so that it is first changing in the first part and the change in that part precedes the change in the whole. If it changes in neither part, it does not change in the whole which is composed of them either. If it changes in one of them, the supposition that it first changes in the whole has been refuted.

Yaḥyâ:
He assumes that AD is the first instant at which the changing subject is changing and that this is the instant at which the preceding rest, which exists in time CA, has ended. Then he says that if it is at rest in the time CA, then in A it must simultaneously be at rest and moving, because A is the end of the time CA, in which it is at rest, and the beginning of the time AD in which it begins to change.

Someone may say: 'How can he say that they must conclude that it is moving and at rest at the same instant, which is the end of the first time and the beginning of the second time, when he has explained before that it is not possible for a thing to be at rest at one instant?'

The solution is that he lets them conclude this in accordance with what they assume, for if they assume that the changing object is changing at an instant, then with greater reason it may be at rest at an instant. Moreover, he may have said 'at rest in A' instead of saying 'not changing in A', since A is the end of the time during which it is not changing, whereas on the other hand it is changing in it, as it is the beginning of the time in which the change begins.

236a19 For it is at rest in A and it has moved in D.

Yaḥyâ:[225]
As he has taken AD as indivisible, it is thought that by saying 'it has changed in D' he means that it [AD] is indivisible.[226] He should have said that it also has moved in A, in order to make clear the contradiction.

236a20 Since AD is not without parts, it must be divisible.

Yaḥyâ:[227]
He means that if it has moved in the whole of the first time, and this first time is divisible, then it must have moved in all its parts. In that case, there must be something which precedes this first time.

236a26 It is clear, then, that there is no first instant in which the object has been moving, because the divisions are infinite.

Yaḥyâ:[228]
He says that if it has been refuted that this supposed instant at which the change first takes place is indivisible, and if it has been established that it is infinitely divisible, so that there does not exist a smallest time, then something in which the first change of the changing thing takes place does not exist at all.

{236a27-36 There is no part of what is changing which changes first.}

Yaḥyâ and Abû 'Alî:[229]
678,9 There are three things existing in relation to changes: the time in which the change occurs, the changing object itself, and the form in which the change occurs. It has been refuted that there could be a first time in which what changes is changing and he will refute the same [thesis] for the form. Now he refutes the claim that there could be some part of what changes which changes first as follows: If some part of what changes was changing first, then it would change in a certain time. Because every changing object is divisible and every time is divisible, half of that part will change in half the time, one fourth of it in one fourth of the time, and so on *ad infinitum*.

This argument of Aristotle applies to changes which occur part by part, not to instantaneous changes; so it does not hold for the instantaneous coagulation of milk.

Someone may say that it does not follow that what changes is divisible according to a division of time, for if a clod of earth moves a certain distance in one hour, it does not follow that half of it moves [that distance] in half the time.[230] The solution is that, indeed, this does not follow, if the clod is actually divided into halves, so that one half is moving and one half is at rest. But supposing that the whole clod moves half the time [covering half the distance], half of it moves half the distance covered by the whole in half of one hour's time.

[Aristotle 6.6]

{237a19-237b22 There is no first instant of having begun to change.}

Yaḥyâ:
691,8 Time, being continuous, may be divided by instants into an infinite number of parts. If the instants are not contiguous, it is clear that

there must a stretch of time between each pair of instants. If change is continuous, then before each change there must be a preceding time in which the changing object has changed.

237a28 Furthermore, what has been said may be explained more clearly in the case of magnitude, because the magnitude along which what is changing changes is continuous.

Yaḥyâ:[231]

He intends to make clear the proposition under discussion in the case of the form [along which the motion occurs]. This form may be quantity or quality. He starts to explain it in the case of quantity, i.e. place. His argument, saying that it is continuous, is not different in the cases of place and of time, for time is also continuous. There is a difference, however, between the cases of place and quality, because quality is not continuous by itself, but only by accident.

237b8 ... just as lines may be infinitely divided so that one part is always increasing and the other decreasing.

Yaḥyâ:

He means that if as one divides a line into two halves, then divides one of the halves into two halves and adds one of them to the other half [of the first division], so that one of the halves keeps growing with the parts we have cut off from the other, this process of division continues without limit, then something similar holds for time.

237b11 But this does not always hold for the thing itself which is being generated, but sometimes for something else, i.e. a part of that thing, such as the foundation of a house. 692,8

Yaḥyâ:[232]

He means that if what is generated must have been generated before, then it is not necessarily the thing which has finally come to be which has been generated before, but something else. For example what has come to be a man has come to be before, but then it was not coming to be a man, but flesh or something other than man. Similarly, one may say about the generation of a house that the foundation was generated before that, and again its parts before the foundation as a whole.

He has added the words 'sometimes it is something else', because for some things the thing which is generated is the same as what was generated before. This is the case for homeomerous things which change by means of growing. For flesh and other homeomerous things may be generated out of what is not flesh, but blood; but we may also find that flesh is growing out of flesh, i.e. flesh as such. Then its growth occurs out of flesh, and the preceding flesh is not identical to the grown flesh, but they are both flesh as such.

Yaḥyâ:[233]

696n.1 Some people have taken the word 'rest' here [238a21] as standing for 'coming to rest'.

[Aristotle 6.7]

> {237b23-238a31 It is not possible for a finite motion to require an infinite time, nor for an infinite motion to be accomplished in finite time.}

Yaḥyâ:

698,12 He has solved Zeno's problem,[234] – that if motion existed, then the infinite would have to be traversed in a finite time, because the number of points on a magnitude is infinite – by saying that the infinite may be infinite in respect of division; this kind of infinite exists in magnitudes which are finite and it is not unacceptable[235] to traverse what is infinitely divisible in a finite time. The impossibility is that what is infinite in extension should be traversed in a finite time. Now Aristotle is going to prove this impossibility.

He first establishes the validity of two principles, the one that every motion takes place in a time, the other that in a longer time a moving thing covers a greater magnitude.

Yaḥyâ:

Suppose something traversed a finite magnitude in an infinite time; then in a finite time a part of that magnitude would be traversed. This part would be a measure of the whole, because the whole is finite. It is not possible for this part to be traversed in an infinite time too, as the whole is traversed in a finite time; for what is due to be traversed in an infinite time cannot be traversed at all; for the infinite has no end. The same could be said of coming to rest.[236]

> **237b28** For if we take a part of the motion which is a measure
> 699,1 of the whole, then the whole motion is completed in equal periods of time, the number of which is the number of the parts of the motion.

Yaḥyâ:[237]

He means that you may take one cubit (*pêkhus* – *d̲irâʿ*) and then cut a part from it, such as an inch (*daktulos* – *iṣbaʿ*). The phrase 'the number of which is the number of the parts of the motion' means that if the whole is divided – and evidently the number of inches is twenty-four – then the number of periods of time is twenty-four.

238a1[238] If one part of the moving object must have moved before another part ...

Yaḥyâ:
He says this because the motion is not a motion of something which moves instantaneously, so if the stretch is one cubit, then one inch of it has moved before another.

238a2 This is evident, that in earlier and later periods of time one part is traversed after the other.

Yaḥyâ:[239]
He refers either to the moving object[240] or to the stretch along which the motion occurs. Then part of the motion occurs in part of the time, and another part in another part of the time.

238a4 For as the time becomes longer a different part of the motion will always be completed in it.

Yaḥyâ:[241]
He says that a part of the moving thing or of the motion moves in an earlier time and another part in a later time, so that one part moves after another.

Yaḥyâ:
If every change is in a time, whether it is a local motion or any other change, then the change passes together with the time, and both run parallel. Therefore one of them cannot be finite while the other is infinite.

{238a32-238b22 Finite magnitude cannot cover an infinity in finite time, nor can the infinite traverse the finite.}

Yaḥyâ:[242]
In his phrase [238b11]: '... either by moving along it, or by measuring 703,13
it off' Aristotle means by 'moving': walking or something like that, and by 'measuring off' that it is passed over, as the cubit passes over the distance when it is being measured.

Yaḥyâ:[243]
If it has been proved that the infinite and the finite cannot traverse an infinite magnitude in a finite time, it is also proved that they

cannot have an infinite motion in a finite time. Similarly, if it has been proved that they cannot have an infinite motion in a finite time, it is proved that they cannot cover an infinite distance in a finite time, for motion takes place along a magnitude (path), as every local[244] motion occurs in a place.

[Aristotle 6.8]

{239a23-239b4 What is moving can be at a position only at an instant, not over a period of time.}

Yaḥyâ:
712,19 If what changes with all its parts is comprised in a whole period of time [while it remains in the same condition], and if every time is divisible, then it must exist in all parts of the time in the same condition, and this is the characteristic of rest.

Then he [i.e. Yaḥyâ] says: What changes, however, from black to white must pass over an infinite number of intermediate colours. For the intermediate stages between black and white are infinite in number, because just as time and motion are [infinitely] divisible, this holds too for the intermediate colours. Therefore what changes passes over an infinite number of things.

If someone were to say that this division is potentially, not actually
713,1 infinite, in the same way as the time, we say that if these intermediate stages are potential and not actually passed over, it follows that those which are actual are passed over, and that what is passed over is actually finite. As time is infinitely divisible, the changing object must be in each of the intermediate states during a certain time, and this makes it impossible for motion to be continuous. If we disallow this, we must disallow that what changes would be in an intermediate state during a certain time.

[Aristotle 6.9]

{239b5-9 Zeno's paradox of the arrow depends upon the mistaken assumption that time is composed of instants.}

Yaḥyâ:[245]
711n.1 The phrase [239b6] ' … and if everything which moves locally is always in a now' should be supplemented by: 'and everything which is in a now occupies a space which is equal to it.'

Yaḥyâ:
711n.3 [The space is] equal to it, when it is in a space without moving, for the now is not divisible, and the space is not larger; so it is at rest,

and then the arrow is resting and in motion, because in a now it is resting; this is the problem.

{239b9-240a18 Zeno's other paradoxes of motion: the Dichotomy, Achilles and the Stadium.}

Yaḥyâ:[246]
[239b33-240a18, the Stadium] He does not assume one row of bodies (*stadion* – *mîdân*), but two, one of them referred to as AA and the other as BB,[247] but because they are equal, he does not [explicitly] mention that there are two rows of bodies and therefore his discussion becomes unclear. He says that one [of the two moving bodies] moves from the middle of the racecourse and the other from the end; he does not refer to one and the same row of bodies, but he means that one of them moves from the middle of the non-moving row of bodies and the other from the end of the moving row of bodies. 715n.2

239b24 But this problem [i.e. the Achilles paradox] goes further in that it says that, as the story goes, even the quicker will not succeed in his aim of overtaking the slower. 722,6

He says:[248]
The cause of the two problems [Dichotomy and Achilles] is one and the same, i.e. the infinite division of the magnitude, but the manner of division is different. In the first problem the magnitude is divided into two equal halves, in the second one the division must be into unequal parts, in accordance with the different [velocities] of the motions, i.e. the quicker and the slower.

Yaḥyâ:
The fourth argument brought forward by Zeno, [the Stadium], runs as follows: If motion existed, then the same time will be equal to double the time and equal to half the time from the same point of view. Let us suppose a stationary row of six equal quantities denoted by six A's, and let us further suppose a row equal to it, with one of its ends opposite the middle of the A's, consisting of six quantities equal to the six A's, denoted by six B's. And let us suppose a row consisting of six equal quantities equal to the six B's, denoted by six C's, with one of its ends also opposite the middle of the row of A's, but at the other side in relation to B. Suppose the rows of B's and C's move towards each other with equal velocity, so that they move in opposite directions.[249] Then the row of B's passes the three A's which it had not yet passed – they are the A's opposite the row of C's – and in the same time the row of C's passes the three A's which were opposite the row of B's. In the same time the row of six C's has passed the row of 723,1

six B's, and because each of the six B's is equal to each of the six A's, the time of passing them must be twice the time needed to pass the three A's, if the velocities of the motions are equal, because they [i.e. the three A's] are half the number of those quantities [i.e. the six B's or A's]. As we know that the row of C's has passed the six B's in the same time as the three A's, it follows that the same time is equal to its double and that it is one half of that double, which is a contradiction.

The mistake in this argument is that Zeno supposes that what moves along another moving object does so in the same time as it would move along that object at rest, which is not true. This has been explained before.

[Aristotle 6.10]

{240b8-241a26 Nothing which lacks parts can move except accidentally.}

729,5 **240b13** The motion of the parts may be different etc.

Yaḥyâ:[250]
Having explained that something may move by accident, like a part in a whole, as wood is part of a ship, or like a passenger on a ship, or like the blackness of a ship, he now wants to explain the difference between the motion of a part and the motion of the blackness or the passenger. He explains that the blackness or a passenger of a ship does not contribute to the motion of the ship, although they accidentally move with it, whereas the motion of a part contributes to the motion of the whole, because the motion of the whole is composed of the motion of the parts.

The motions of the parts may be different from one another. For what is far from the axis of a sphere moves faster because of the extension of the circle, and what is close to the axis moves more slowly, because what is close to what is at rest almost resembles it.

The motion of the indivisible in a magnitude is not the same as the motion of the parts in a whole, because it does not contribute anything to the motion of the whole.

{241a26-241b20 Change cannot be infinite, at least when it is between contraries.}

Yaḥyâ [ad 241b2: Local motion will not be finite in this way]:[251] 730n.1
He means that local motion is not [always] finite in this sense, i.e. that it occurs between contraries. For circular motion does not occur between contraries, nor does the horizontal (sideways) motion of animals.

[Aristotle 7.1]

{241b24-243a2[252] If something is in motion, it must be kept in motion by something else (version Beta). To prevent an infinite regress of movers, there must be a first one.}

(ad 241b27: [Let AB be something which moves by itself,] not in virtue of the fact that something belonging to it is moving):[253]
As we say that a sleeping person has moved if he has moved his foot or hand. 734n.1

Yaḥyâ:
Aristotle presents two principles in a concise way. One of them is that 743,2 if something moves, not in virtue of something else, then it is not necessary that its motion stops when something else stops moving. The second principle is derived from the first one by means of 'conversion by antithesis' (*al-'aks 'alâ sabîl at-taḍâdd – hê sun antithesei antistrophê*)[254] i.e. the opposite of the predicate is stated, and then the opposite of the subject follows. The resulting proposition is that if something stops moving because something else stops moving, then it will not move by itself.

He also summarily mentions that everything that moves is divisible. Then he explains what he will discuss here, namely that what is moving by itself, not in virtue of a part, has a source of motion. He denotes the thing that is moving by itself by AB and he lets it be divided at C. If AC is at rest, then the whole AB must be at rest. For if that were not the case, so that it would be moving, then it would be moving on account of a part, not by itself, and it was assumed that it was moving by itself. Now, as the whole stops moving when a part stops moving, its motion occurs by something else. This is a dialectical method of demonstration.

241b27 I say in the first place, that if we think that AB is <not>[255] moving on account of something …

Yaḥyâ and Abû l-Faraj:
He means that if we think that when Zayd is moving, he is the mover as well as what is moved and that there is no soul in him which moves

him, then this would be something similar to our thinking that the wax[256] is moving, without an animal in it to move it.

Yaḥyâ:
If every mover were to be in motion, while it is established that everything which is in motion has a mover, it would follow that there would be an infinite number of moving things and that they would move at the same time, because the motion of one is the cause of motion of another. If that were the case, it would follow that there are an infinite number of motions in a finite time.

Then Aristotle starts to show the impossibility of this, saying that it has been shown before that it is impossible for a motion which is numerically one to be infinite in a finite time. The existence of an infinite number of motions in a finite time has not been shown to be impossible. And it is not impossible that many motions exist in the
744,1 same time. Aristotle establishes the proposition in a way resembling a demonstration, saying that what is moved and what causes the motion, i.e. the effective mover, are either continuous – as when they form a mixture, such as the heaviness that is in us and moves the water[257] in us downwards – or they are touching, so that their extremes coincide, for they must be in contact. In both cases these motions will form together one infinite motion.

We may give a concise proof as follows: if we assume the situation is as described above, it follows that there is an infinite number of moving things, and the infinite cannot actually exist.

242a16 Also what moves locally must be moved by something else.[258]

Yaḥyâ and Abû l-Faraj:
He brings the discussion onto local motion, and says that motion is passed on from one thing to another. He discusses local motion because it is the most general kind of change[k], as every other change[k] exists only if local motion exists. By 'other changes[k]' he means growth and diminution, and also alteration, because an alteration must always be preceded by a motion of what causes the alteration towards what is going to alter; these must approach one another. Thus, what is said of local motion is also valid for the other changes[k], because local motion is the most general change[k].

242a29 For although each of them is moved by another, yet the motion of each may be taken as numerically one.

Yaḥyâ:[259]
He says that although each one of the moving bodies moves another,

this does not prevent each from having its own motion. Then he will say that if these motions are many – they are infinite [according to what was assumed] – and when they are composed to form one motion, this must be one motion.

Yaḥyâ [ad 242b7]:[260]
Good and bad exist in all categories: in substance, in quantity, in quality, and in the others. Therefore he says that motion from good to bad is not one in species. 745,1

242b17 For their motions may be either equal to the motion of A, or greater.

Yaḥyâ:
He says that the motions of all the moving bodies are either equal or unequal. Whatever they are, in either case they are infinite. It is possible to interpret equal and unequal here either as referring to a difference in velocity, or to a difference in size of the moving objects. The latter is more suitable, because if the motions were different in velocity, then what is moving will be disturbed, or there will be no corporeal motion.[261]

242b30 It makes no difference, whether they are finite or infinite.

Yaḥyâ:
He means that it makes no difference for what we have said whether the body which is formed out of the bodies is finite or infinite. When the number of motions is infinite, then the motion which is composed of these is infinite, for the magnitude which is composed of an infinite number of magnitudes must be infinite, and similarly the motion composed of an infinite number of motions must be an infinite motion.

Alpha **242b67** For if what is composed of A, B, C, D is infinite …

Yaḥyâ:
Regarding this he has said before that it makes no difference whether it is finite or infinite. He now takes it to be infinite; this is the case which applies here.

242b34 The argument is not defective because it uses the assumption which was posited.

Yaḥyâ:
He wants to make clear that this explanation is not faulty, although it proceeds by deriving a contradiction (*reductio ad impossibile* – *ḵilf*); there is no difference between a *reductio ad impossibile* and a direct proof, for the impossibility follows from what was posited, and what was posited is not impossible.

[Aristotle 7.2]

{243a3-245b2 Agent and patient must be in contact. (a) The case of local motion, (b) quantitative change.}

Yaḥyâ:
753,15 What has local motion moves either by itself, or by something external. As for what moves by itself, the situation is clear: its mover is in contact with it and there is nothing between them. Examples are soul and body, heaviness and a stone, lightness and fire. As for what moves by something else, there are two cases: what accidentally moves and what essentially moves. What accidentally moves is what is carried, and this may occur in three ways: it is carried on something else, it is carried on water, as for example a traveller on a ship, or it is carried in the air, as a letter is carried by a bird. What essentially moves, while the cause of its motion is in something else, may do so in two ways: either it changes its place as a whole, or its parts change their places. The latter case is turning. As for what changes its place as a whole, the moving cause either moves it away from itself, which is called pushing and moving away, or he moves it towards himself or towards something else, but not towards something opposite himself, which is called pulling. Pulling towards something else is called
754,1 bringing closer and bringing together. As for pushing, if the pusher pushes it and remains attached to what he pushes, then it is called pushing on (*sawq* – *epôsis*), if he does not remain attached to it, it is called pushing off (*zajj* – *apôsis*).

Furthermore, if the mover of something brings a motion into it which is stronger than its natural motion, either in the opposite direction or in any direction, i.e. the direction of the thing, then this is called throwing (*ramy* – *rhipsis*). If this is not the case, then it is not called throwing: if I let a stone fall, I do not just say that I have thrown it.

All these motions may be reduced to two kinds, combination and separation, and they all need a mover which is together with them, i.e. which is in contact with them, without something in between. The mover which is the goal [of the motion] does not need to be in contact with what is in motion. That which is in contact with it is what causes motion as an efficient cause.

The translations of Isḥâq and ad-Dimašqî have [Beta **243b29**]: 13
'Every local motion is combination and separation.' The translation of ad-Dimašqî adds [Alpha **243b9**]: 'except those involved in generation and perishing.'

Yaḥyâ:
When combination and separation are reduced to pushing and pulling, he makes an exception for generation and perishing, for, indeed, generation is associated with combination and perishing with separation, but generation does not mean the same as combination nor perishing the same as separation.

Yaḥyâ:
If turning and other kinds of motion may be reduced to pulling and pushing, then what about a thing that moves to its natural place while a piece of wood attached to it is moving together with it? Do we say that this motion occurs by pulling or pushing?

We do not say about wood which is floating on water that it is carried, because something is carried only when it is not at its natural place. Wood is floating on water on account of the air in it and air is at its natural place when it is above water.[262]

Water is neither pulled nor pushed when it is moving to its natural place. Something is pulled or pushed only when it is moving away from its natural place.

A thing is carried, if it is moved to a place other[263] than its natural 755,1
one. Thus, when wood is floating on water, it does so on account of the parts of air in it. If something is moving downwards together with water, it does so on account of the earthy parts in it; the downward motion of the earthy parts does not occur by pulling or pushing, but these parts pull the air with them, which exists in between them.

Alpha **244a7** This is clear from the definitions we have given on these matters.

Yaḥyâ:
He means that what we have maintained, namely that the mover which pushes and pulls must be in contact with what is pushed and pulled, is clear from the definitions of pulling and pushing we have given.

In the Syriac version and the version of ad-Dimašqî: Aristotle [Alpha **244a9**]: ... when the motion of what pulls is quicker than the motion which would separate from one another the things that are continuous [i.e. what is pulling and what is pulled].

Yaḥyâ:
Or: what is disposed to be continuous, such as when Alexander wants to separate himself from the Greeks, whereas he is disposed to be in contact with them. If that which pulls him to an agreement with them is stronger than what tries to separate him from them, the pulling is effected, as it is quicker and stronger than the motion of separation.

In the version of ad-Dimašqî: Aristotle (Alpha **244a11**): One might think that there is a kind of pulling which occurs in another way: that wood pulls [fire] in a way different from what was described here.

Yaḥyâ:
He says that not everything which pulls is in contact with and together with what is being pulled. For burning wood pulls the fire which burns it towards itself and prevents the fire from moving up.

This is not a correct example, because the fire does not become attached to the wood, but the wood is turned into smoke which is pulled away. How could that which is not attached be pulled? Furthermore, the wood gradually perishes, so how could it pull? Correct examples include the magnet which pulls iron to itself, and amber which pulls dust particles to itself. These are movers which are not in motion.

In the version of ad-Dimašqî: Aristotle (Alpha **244a13**): It makes
756,1 no difference whether that which is pulling is in motion or is stationary when it is pulling; in the latter case it pulls to the place where it is; in the former case it pulls to where it was.

Yaḥyâ:
That which is pulling must be in contact with what is pulled, whether it is at rest or is in motion with what is pulled. Thus, the iron must be in contact with the magnet which pulls it to its place. If what pulls is moving, then it pulls what is pulled to where the pulling cause was. For the man who wins over (pulls) Alexander towards peace with the Greeks, namely Menelaus, pulls him to where he (Menelaus) was. If that which pulls is at rest, then it pulls what is pulled to where it is.[264]

Furthermore, the magnet pulls because particles are issued from it which penetrate the iron so that it is pulled [to the magnet]. [Another explanation is] that the air is changed so that it gets a power to pull; then that which pulls is the air which is in contact with the stone.

Yaḥyâ:
Having discussed local motion and having explained that its effective cause must be in contact with [the moving body], he explains that this must also be the case for alteration. He says that alteration occurs in the third kind of quality, namely the affective qualities or affections (*hai pathêtikai poiotêtes – al-kayfiyyât al-infi'âlât*).[265]

Animate bodies as well as inanimate bodies change by the influence of sensible characteristics; vision[266] is influenced by the sensible characteristics only in a complete way.

When bodies are generated they are altered by colours. The sense of smell is influenced by odours,[267] the sense of taste is influenced by flavours and the sense of hearing is influenced by sounds. It should be noted that bodies which break when an enormous sound occurs do not break because of the sound, but because of the motion of the air. The sense of touch is influenced by touchable characteristics, i.e. roughness and softness, and by heat and coldness. It follows from all this that the last cause of alteration is in contact with the first thing which is altered.
He says this because the colours, sounds and odours reach the senses 757,1
only by means of something intermediate between them and the senses. What causes the alteration of the senses is the air which receives these influences, and the air is in contact with these senses.

244a28 The sensible characteristics are those by which bodies differ from each other.

Yaḥyâ:
By sensible characteristics he means colours, flavours, heaviness and lightness, wetness and dryness and such things, for they are sensible, and it is by these that bodies differ from each other and it is by these that the senses are altered. One should not understand by the expression 'sensible characteristics' those which are common for two senses, such as the shapes themselves, nor what is sensible by accident, i.e. substance.

244b27 The animate may be altered by everything by which the inanimate may be altered, but the inanimate may not be altered by everything by which the animate may be altered.

Yaḥyâ:
Everything by which the inanimate may be altered, such as heat and coldness, may also cause an alteration of the animate, because the animate may also alter by heat and coldness. The animate may also alter by things which cannot cause an alteration in the inanimate, because the animate may change and be influenced by the sensible characteristics when the sense-organs come in contact with them and

perceive them. The inanimate things are not changed in this way. If they are changed by them, as a mirror is changed by colours, they are not aware of it.

Animate[268] bodies are sometimes not aware of a change which occurs to them either, but this only occurs when the change involves an insensible part, such as when our hair becomes grey.

[Aristotle 7.3]

{245b3-248a9 Change of quality (alteration) is change of *sensible* quality. Apparent exceptions explained.}

Yaḥyâ:

766,12 He explains that there is no alteration in shape and form, although one might think that alteration occurs in these things. For what has changed its shape, such as copper or wood, has not changed its substance.

He also shows that there is no alteration in dispositions or conditions. He does not show that there is no alteration in capacity and incapacity, because this is obvious. For alteration occurs in the accidents of a thing, its substance remaining unchanged. Sperm which has the capacity to become a rational being and sperm which does not have that capacity differ in substance. Similarly, when a child lacks the capacity to learn – capacity here means a suitable disposition – and then acquires this capacity when he gets stronger and becomes a young man, then this is a change in substance.

245b19 Alteration occurs only in those things which may essentially receive an influence by them [i.e. by the sensible characteristics].

Yaḥyâ:

The things which are influenced by the sensible characteristics are the bodies whenever they are being generated, or become hot, or dry, or are influenced by other affections. The senses are influenced while they are receiving their forms, in accordance with what is said in *On the Soul*.

Yaḥyâ:

Just as we do not say of a house that it has altered, but rather that
767,1 it has come into being, we do not say that something has altered to a statue, but that it has come into being. If nothing has changed of the substance of the house and the statue, [then it is an alteration].[269]

Yaḥyâ:
Coming to be is not alteration, but it necessarily involves alteration, because matter is condensed or rarefied, or undergoes some similar affection. These are alterations.

Yaḥyâ:
There is no alteration in states, i.e. in conditions and dispositions, because states are either excellences or defects, occurring either in the body or in the soul. The excellences of the body and of the soul are all perfections. One does not say that the acquisition of a perfection by something is an alteration, but a perfection is said to have come into being as the thing has received its properties in the most perfect and laudable form. Therefore we do not say that a thing is a horse if it lacks one of the things by which a horse is a horse. When all the things by which a horse is a horse come together [to form a horse], then this is not called an alteration, but one says that a horse has come into being. An example is health, one of the corporeal excellences, and healthiness of the soul, one of the excellences of the soul. Illness is a corruption, as health is a perfection.

Furthermore, when states are excellences and defects, and corporeal excellences consist in the right balance, and the same holds for excellences of the soul, then there must some balance in them [sc. in body and soul]. This must be a balance of things which are in balance, and this is a kind of relation, so excellences are relatives. Defects are also relatives, because they consist in a being out of balance of what could be in balance.[270] There is no alteration in relatives, because a relative is a form, and a form is acquired instantaneously. The acquisition of a form is not an alteration, nor is coming to be, since alteration is a motion and a process.

This may be explained as follows: I come to the right of Zayd after having been to his left, when Zayd moves. This [relation of being at his right] is acquired by means of a motion, for Zayd moves from one place to another, and at the same time it occurs that I am at his right or left. Indeed, this [motion] is not something which takes place for the relative itself, but it occurs to something else. Thus it [sc. the relative] is not subject to a motion, but the motion occurs in something else and the relation is acquired as a result of it. Similarly, also when the elements move, then being in balance and being out of balance will follow as a result of the motion.

246b21 ... either internally in relation to one another, or in relation to the surrounding atmosphere. 768,1

Yaḥyâ:
By 'internally' he means health which arises from a balance of the

temperaments. By ‘in relation to the surrounding atmosphere’ he means being in balance with the atmosphere, for if the temperaments are in balance, they are not necessarily in balance with the atmosphere, which might make health perfect.

> Alpha **246b14** But it may be that their becoming and perishing is necessarily the result of the alteration of something else, just as in the case of form and nature.

Yaḥyâ:
He says that just as form and nature may come into being and cease to be, not by an alteration which occurs to the thing itself, but by an alteration which occurs to something else – e.g. by an alteration of the substrate from cold to hot – in the same way health arises as a result of an alteration which primarily occurs to the elements, so that they come into balance. Then, a balance is formed in a secondary way in the members of the body and in the body as a whole.

Yaḥyâ:
Having explained that there is no alteration in corporeal states, Aristotle continues the discussion, saying that it does not exist in states of the soul either.

The states of the soul may be divided into moral states and intellectual states. Intellectual states exist either potentially, such as the potentiality existing in children, or actually.

Moral states are relatives, because those which are excellences consist in a balance as regards receiving pleasure and pain, and those which are defects consist in a being out of balance as regards receiving pleasure and pain. We have said that balance is a relative, because it is a balance of what is in balance.

> In the Syriac version: Aristotle [Alpha **247a3**]: Furthermore, excellence has a good influence on someone to meet his proper affections, and defect has a bad influence.

Yaḥyâ:
[By proper affections] he means the laudable affections of pleasure and pain.[271] By affections he means the affections of the elements.[272]

769,1 Moral excellences and defects do not come into being by an alteration in them, but by an alteration in something else, namely the senses. He has shown this in the book *On the Soul*. That the moral excellences come into being by an alteration in the senses is shown as follows: The moral excellences occur with regard to pleasure and pain. Pleasure either exists at the present time, whenever our senses experience it, or in the past time, whenever we remember what our senses have experienced in the past and this gives us pleasure, or in the future, whenever we hope to experience something similar to what our senses have experienced. All this goes back to the senses. Thus, when the senses are altered by some sense-object, they move the imagination, and the imagination moves the desire and the nerves, and by this pleasure and pain arise.

If the soul moves towards a sense-object, not insofar as it is that sense-object, it is a vice, such as when the soul at the sight of a visible object moves towards it, not insofar as it is visible, but insofar as it is tangible. When it moves towards the sense-object insofar as it is that sense-object, it is not a vice. For example, when two men look at a visible object, and one of them is affected insofar as it is visible, whereas the other is affected and moves towards it insofar as it is tangible, then that is a vice.

Alpha **247b1** There is no alteration in the states of the intellectual part of the soul either.

Yaḥyâ:

Having explained that the moral states are not alterations, and are not acquired by alteration, he now starts to discuss the states of the intellectual part of the soul, i.e. knowledge. He says that the possession of knowledge is even more truly said to be a relative than the moral states. Knowledge and the possessor of knowledge both have a relation to what is known. The possessor of knowledge and knowledge do not form a relation, for if the possessor of knowledge and knowledge necessarily had a relation – as the possessor of knowledge is knowing because of the knowledge – then it would follow that everything which receives something would be a relative. Then what is hot would be a relative, because it is hot on account of the heat. If a relative is not an alteration, and is not acquired by an alteration, then this is true also for the states of the intellectual part. The possessor of knowledge does not acquire this state after having not been a possessor of knowledge by means of an alteration. For if someone receives knowledge, while having the capacity to know, then what occurs is not that his capacity moves towards knowledge, but that the soul is provided with knowledge, when something else appears to it. For example,[273] when the soul gets to know that the white of Zayd scatters the visual rays, and that the white of ʻAmr and of K̲âlid and of others do the same, then he knows the universal thing, namely that white scatters the visual rays, and then he may establish the same property for the white in China and India.

Then Aristotle[274] turns from the states to the actualisations of the intellect and he says that they are not an alteration, and are not acquired by an alteration. For alteration must be a transition from something potential to something actual. Actuality does not arise by an alteration. Similarly, touching and seeing are not alterations or motions. The touchable bodies are moving, and this will be a local motion; when they touch us this motion stops. Similarly, when we open our eyes and this local motion has stopped, we are able to perceive and see.

Alpha **247b9** The original acquisition of knowledge is not a becoming or an alteration.

Yaḥyâ:
Having explained that there is no alteration in the intellectual states of the soul, nor in their actualisations, he now starts to explain that there is no alteration in the capacity and disposition for them either.

By acquisition[275] of knowledge he means here the acquisition of the disposition to receive knowledge actually, such as what young men have, but children do not have, for a child does not have the disposition (capacity). Therefore someone cannot have the capacity for knowledge as long as he is a child. This capacity belongs to the second of the kinds of qualities,[276] which we said Aristotle has not discussed, because it was obvious that there could be no alteration in them. The capacity discussed here is the disposition [for knowledge], which exists in young men. He says that it does not arise by an alteration. It could exist in a child if it were not diverted and impeded from it by its restlessness and the motion of the natural forces. When this restlessness subsides he becomes able to know. He explains that this is not an alteration. The explanation of all this is that a child is fiercely moved by forces of nourishment, change and growth, because nature wants to bring it to completion.

Thus by strong motions and restlessness the capacity is impeded, just as restlessness in drunken and sleeping persons prevents them from actually exercising their knowledge. Just as someone who
771,1 awakes and then knows and acts accordingly, not because he becomes knowing, but because the restlessness disappears from him and things become clear, in the same way the disposition a young man gets arises when the impediments of restlessness disappear.

Alpha **247b10** For a man becomes knowing and understanding when his soul arrives at a state of rest and comes to a standstill, and there is no coming to be which leads to a rest.

Yaḥyâ:
He means that a man gets knowledge when his essence is at rest and stable and the restlessness and natural motions have been expelled, and rest does not arise by a becoming or an alteration. It is not necessary that every motion proceeds to rest and that one of the

contraries [sc. motion and rest] is a process towards the other or has a becoming. For rest is not a process to motion.

The situation of children is similar to the one of a man who sleeps or is drunken and has the potentiality to be knowing, but does not know because of the restlessness. Children may become knowing when their restlessness disappears, either by their own nature, or by something else, namely instruction. What is meant by 'by their own nature' are the things which are known by intuition (*badîhat al-'aql*).

Alpha **248a3** In both cases this occurs by an alteration of something in the body.

Yaḥyâ:
By 'both cases' he means the things in which nature is sufficient [to bring them about], and the things which are brought about by instruction. For these things an alteration is necessary which comes about in the body, as we say that the body is altered by the motions of the natural forces and then comes to rest and also as the sense-organs are altered by the sensible things and we must perceive in order to know. Furthermore, those who frequently attend lectures on the disciplines of knowledge get lean and dry bodies, which results in their <not>[277] easily becoming annoyed. Also, if there were not those alterations and affections connected with the body, we would not be 772,1
able to explain the expressions in the face of someone showing that he has understood what we say and the other expressions showing that he has not understood us.

Alpha **248a6** It is clear from what has been said that alteration and being altered occur in sensible things.

Yaḥyâ:
By 'alteration' he means the form itself, and 'being altered' is the affection of that which is subject to alteration by the form itself.

[Aristotle 7.4]

{248a10-249a29 Under what conditions can changes be compared with each other? Rectilinear and circular motion.}

Yaḥyâ [ad 249a2]:
The name [of the change] must be synonymous[278] in species and the 779n.1
subject [that receives the change] must be one and the same, i.e. the proximate subject, not the surface only, but the body which has its own specific constitution. For a body does not receive any arbitrary attribute.

Yaḥyâ [ad 248a12]:

783,8 Aristotle means that a circle can be equal to a straight line if one accepts the view that a motion along a straight line can be equal to or comparable to a motion along a circle.

Someone might say: 'Is it not true that an arc [of a circle] is greater than the chord and that a small part of the arc having a size of a cubit is smaller than the chord when the chord is ten cubits? Thus, if a straight line and a circular line can be smaller and greater compared to one another, then equality must also be possible, because the greater cannot shift to the smaller without having passed what is equal.' The solution is that we say that the chord is greater than a part of the arc or that the arc is greater than its chord, if we have stretched out the arc. If we stretch out an arc it becomes straight and it is not curved any more. As long as it has its curved form, it cannot be equal to something straight. Things which are not similar are not comparable.

Furthermore, it is not necessary that if things can be larger and smaller than others, they can be equal too. For the sharp angle formed by a straight line and one half of a circle is larger than any sharp angle, and the sharp angle between a straight line and the convex curvature of an arc is smaller than any sharp angle.[279] There is no sharp angle formed by a straight line and a circular line which is equal to a sharp angle formed by two straight lines, for if that were the case, then a straight line would coincide with a circular line. If we move a line [turning] on the [intersection of a circle and its] diameter, we produce an angle formed by a straight line and one half of the circle – that is the largest angle – and an angle formed by a straight line and the convex curvature – that is the smallest angle. These angles are produced one after the other without an angle being produced[280] which is equal to the sharp angle formed by the two straight lines [sc. 784,1 the moving line and the diameter].[281] Thus the statement has been refuted that if something shifts from being larger to being smaller, this can occur only if it passes what is equal.

Yaḥyâ:[282]

If two motions, one along a straight line and the other along a circle, are equal, then the two distances, the straight one and the circular one, must be equal, because if the motions are equal then the distances which they cover must be equal too.

Similarly, if a local motion and a qualitative motion were equal, then the quality in which the motion occurs would have to be equal to the length travelled by the local motion. Things have an equal velocity, not when they move during an equal time, but when they move the same distance in an equal time.

248a19 It is absurd to maintain that something could not move along a circle and also along a straight line in a similar way.

Yaḥyâ:
By 'in a similar way' he means with equal velocity, so that it moves along a straight line and along a circle in the same time. He also says that if the motion along the circle were to be quicker for instance, then it could move along a part of it in the same time as the time in which it moves along the straight line. Then there would be something which moves in the same time along a circle which would be equal to a straight line.

Yaḥyâ:
Aristotle discusses this problem, saying that if these two motions were equal, then a straight line would be equal to a circle.

Yaḥyâ:
Aristotle gives us a rule concerning things which are comparable. This rule is that [in order to be comparable] things must be synonymous, not homonymous (equivocal). If one uses the terms 'much' and 'double' for water and air and one means a larger quantity, then these terms for water and air are synonymous, so they are comparable; if one means much and double as regards to the force and the quality, which are specific for each of these things, then they are not synonymous, because 'much' in relation to water refers to density, and in relation to air it refers to rarity. The same holds for the term 'double' in sweetness and sourness, and therefore they are not comparable. 785,1
What is denoted by homonymous names is not comparable, for example sharpness in sound, in vinegar and in a piercer. In this way the above-mentioned rule is verified.

Yaḥyâ [ad 248b22]:
The white in a horse and a dog is not divisible in the sense that it is not further divisible into species, as colour is.

248b10 Maybe the meaning of 'velocity' is not the same here as there.

Yaḥyâ:
After he has said that synonymy means comparability, he says that the velocity of a circular motion is not the same as the velocity of a rectilinear motion. This resembles a similar statement on local motion and alteration. [In the latter case the statement holds] because they are of a different genus. Circular motion and rectilinear motion

are both local motions, but this does not mean that they have a common genus, for they are primary and secondary in that which comprises them and so that is not a genus for them. We shall show that circular motion is primary, and that it can exist without other motions, whereas other motions cannot exist without it.

> **248b21** Someone might say that this is the case [sc. incomparability] because that which primarily receives [the attributes] is different.

Yaḥyâ:
He seems to say that if synonymous attributes exist in primary subjects which are not different, then they are comparable, and if they exist in different subjects, then they are not comparable, such as 'many' in water and air, because these have different natures.

He says: 'in that which primarily receives them' because white may exist in a body or in a surface, but the surface is that which primarily receives it and it exists in the body on account of its existence in the surface.

> **248b23**: Similarly, they are comparable in respect of size.

Yaḥyâ:
He says that just as the white in a dog and in a horse are comparable because its subject is one and the same, namely the surface, a dog and a horse are comparable in respect of size, because the subject [of size] is the same too, namely the prime matter.

786,1 **249a2** Moreover, a subject does not receive any arbitrary attribute, but it primarily receives one specific attribute.

Yaḥyâ:[283]
One subject receives one nature. Surface receives colours, nothing else; wetness receives flavours, nothing else; the vibrancy[284] of the air receives sounds and nothing else, and this holds for the other things in the same way.

> **249a13** What is the reason for this? Is it that local motion is a genus or that a line is a genus?

Yaḥyâ:[285]
He says that it is absurd that an alteration should be equal to a local motion and that a rectilinear motion should be equal to a circular motion. But what is the reason for this? Is it due to the fact that 'motion' applies to transformation as well as local motion, or is it due

to the fact that 'distance along which motion occurs' applies to a circular as well as to a rectilinear distance?

249a15 The time is always indivisible in species.

Yaḥyâ:
It is with good reason that we ascribe the absurdity to these two things, i.e. to the motion and to the line, [both of] which may be either rectilinear or straight, and not to the time. For the species of time cannot be further divided into subspecies. 'Line' may further be divided [into different kinds of line] and 'motion' may also be divided into motions of different kinds.

249a15 We say that this holds at the same time for the motion, namely difference in species. For local motion is differentiated into species because of the specific differentiation of that along which it occurs.

Yaḥyâ:
This is the solution of the problem: We should not ask which of them is divisible [into species], the subject or the form in which the motion takes place. Both of them are divisible. If one of them is divisible, then the other is divisible too, because if the subject is different, then the form in which it occurs is also different, and similarly, if the form is different, then the subjects are different too. The reason is that the proximate subject of something receives one thing only.

Yaḥyâ:[286]
Rectilinear motion is one in species and not further divisible into 787,1
subspecies. Thus the motions of flying and walking are not different in species; they only differ in respect of the kind of instrument.

249a21 Therefore we should investigate what constitutes a specific difference for motion.

Yaḥyâ:[287]
He says that our investigation of the things by which motions differ is useful in order to know when motions are comparable and equal. Motions differ either in genus, as do local motion and alteration, or in species, as do becoming white and becoming black, or in number [i.e. as individual motions], as do the becoming white of a swan[288] and of a horse. Motions which are not the same in species[289] are not comparable.

249a21 This discussion shows that the genus is not a unity.

Yaḥyâ:
He says that from the knowledge of the specific differences for motion we learn that the genus of motion, i.e. motion in the absolute sense, does not comprise one thing.

[ad 249a23 ff.]
The things denoted by homonymous terms may be [completely] different, such as a sea dog and a land dog, or they may be close to one another, like those having a common proximate genus, such as a rectilinear and a circular motion, or they may be close in respect of the relation [they have to something], as when we call the heart as well as a point a 'beginning'.

249a25 When is there a difference of species? Is it when the same attribute is in different subjects or must a different attribute be in a different subject?

Yaḥyâ:
He asks when one knows that the species of two motions are different. Is it when the same thing is in two subjects, or when different things are in different subjects? It is such as we have said just before: it is not possible for one nature to exist in different subjects, but one form is always received by one subject.

{249a29-249b26 The comparability of alterations, generations and perishings.}

796,11 **249a30** We say that if recovery of health is an alteration, and if it is possible that one person recovers quickly and another slowly, then it is also possible that they recover in the same time.

Yaḥyâ:
He assumes here that the recovery of health is an alteration. It has been said that it is a coming to be, which is correct. One should say instead of 'recovery of health': 'becoming white' and 'becoming black'. Then the statement becomes: if it is possible that two things become white, one in a long time and the other in a short time, then it must also be possible for them to become white simultaneously in the same time, or in two equal times.

249b2 But what alteration? [i.e. what aspect of alteration do we mean when we say that alterations are equal?] We cannot speak of equality here; what corresponds to equality in the category of quantity is here 'similarity'.

Yaḥyâ:
He raises a problem, asking whether, when things undergo an equal alteration, it is the subject of the affections, i.e. the body which becomes white or black, [which must be equal] or the affections themselves, like whiteness and blackness. Then he raises another problem, asking how we can say that two alterations are equal, when 797,1
equality can occur in the category of quantity only and alteration does not belong to that category, but to the category of quality.

After that he explains that one may say they are equal and comparable on account of the fact that they occur in the same time. Thus, the term 'equality' applies to them on account of the time, just as the terms 'much' and 'long' apply to whiteness when it resides in a long subject. He makes this clear by saying: 'Let us say that there is equality in velocity when the same change is performed in an equal time' [249b4].

249b5 In what are we to look in order to find comparability, in the subject in which the affection occurs, or in the affection itself? In the case we have been considering here it is because health is one and the same that we can take it [sc. the alteration] as being neither more nor less.

Yaḥyâ:
The statement that the alteration is one may be understood in the way that the subject in which the affection occurs is one, or in the way that the affection itself is one. The truth is that one has to consider both. [One has to ascertain] that the subject is one, the [kind of] alteration is one and the time is one, so that a certain process of becoming white is the same as another process of becoming white in respect of the characteristics we have mentioned.

Aristotle says that what he describes as comparable is the affection itself which resides in the subject and he says that this becoming white is equal to that becoming white.

Yaḥyâ [ad 249b11]:
Knowing that an alteration is not comparable to just any other alteration, we must distinguish the different kinds of alterations in order to know which ones are comparable to which. Aristotle has explained this matter.

249b12 If the things which are moving – i.e. the things that have their motions essentially, not accidentally – are different in species, then their motions are also different in species.

Yaḥyâ:[290]

He says that moving things have affections in respect of which they move, just as motions have affections in respect of which they take place. Just as bodies may be white or black, and are specifically differentiated by these attributes, there are changes[k] in respect of them, and they are specifically different. Just as bodies may be black
798,1 or sweet [and in that case are different in genus], there are changes[k] in respect of these two attributes, and they are different in genus. Just as two individuals[291] may be white, there are changes[k] in respect of the whites of these individuals and they are different in number [i.e. individually].

The words 'not accidentally' should be taken to refer to the affections which belong to the things that change[k], not to the changing[k] things themselves. For whiteness and blackness are different in kind, although the things to which they belong may be of the same species, like Zayd and 'Amr, and this holds for becoming white and becoming black in the same way. These changes[k] may be different by accident, like the whitenings of a man and a horse. These changes[k] are essentially the same in species, but accidentally different, because it is accidental for them that they occur in different subjects.

If two alterations are different in species, such as becoming white and becoming black,[292] and occur in two things which are of the same species, then they are essentially different and accidentally the same, because it is accidental for them that they occur in things of the same species.

Therefore we must not consider the subject, but the affections in respect of which the changes[k] occur in order to know which changes[k] are different in genus, which are different in species, and which are different in number.

249b14 But we must still look into the question how we establish that two alterations are equal in velocity. Should we consider the affection, whether it is one and the same, or similar, in both cases, or should we consider the altered things? ... *until*: We say that they are equal or not equal according to the equality or non-equality of those (i.e. the altered things).

Yaḥyâ:

He wants to explain that one should consider the altered thing, the affection in respect of which the alteration occurs, and the time in which the alteration occurs in order to know whether alterations are one and equal. By the phrase 'so that we should look for instance whether this has become white to this extent and that has become white to that extent' he means the extent of whiteness to which something is white, because the subjects must receive the affection to the same extent and one of them must not receive it to a greater extent than the other, so as to receive more whiteness than the other.

It is also possible that he means that one of the surfaces which have become white is larger than the other. Then the processes of becoming white are one in species, but not equal. 799,1

> **249b19** We should also investigate the same question for generation and perishing, i.e. how are processes of generation equal in velocity?

Yaḥyâ:
Having made clear the conditions under which alterations and local motions are equal and comparable, he now turns to a discussion of the same question for generation and perishing. The conditions under which two processes of generation are comparable, and also two processes of perishing, are the same [as in the cases mentioned above], namely that the time of both processes is the same and that they occur in respect of forms which are of the same species, not of the same genus only. Thus, for instance, the generations of a man and an animal are not comparable, but the generations of one man and another are comparable.

These conditions are also the ones under which processes of growth and diminution are comparable. They are that the increases must be of the same species and occur in the same time, and the same conditions hold in the case of diminution.

> **249b22**: For we have no name to convey this difference, as we say 'not similar' [in the case of alterations]. If substance were number, then we might say 'greater' and 'less' of numbers of the same species.

Yaḥyâ:
He says that we cannot find a name for processes of generation being of equal velocity, i.e. being comparable, nor for their being different, i.e. we cannot find a name for processes of generation which are not comparable – as we say unequal,[293] nor can we find such a name in the case of perishing.[294] We have found such a name in the case of local motion. In that case comparable[295] motions are said to be equal, and not equal when they are not comparable. The same holds for

growth and diminution, because they belong to the category of quantity. As for alteration, this occurs in the category of quality, and to quality belongs similarity; so we say that alterations are similar or not similar. Substance does not belong to quality or quantity, therefore we cannot apply these terms to it.

Aristotle explains this further for the case of substance. He speaks about substance as if it were a number and he does this in accordance with what he wants to say about substance, in order that the discussion will be more clear. He does this because there is no name which
800,1 might be applied to all processes of generation, irrespective of whether they are general or particular,[296] so that beings could easily be divided and it could be made clear which ones are equal and which ones are not equal. Suppose a man consists of five units, another man[297] of three, and still another of seven units. If the man who consists of five units is generated in a certain time and in that time three units are generated in another man, then these two processes of generation are not of equal velocity, because one man is generated in just that time, and the other is not generated in that time. Thus two processes of generation are equal when they occur in the same time in respect of the same species.

Alexander says that he mentions number here following the opinion of Pythagoras and his school, namely that the principles of things are numbers. He further says that it is also possible that he mentions numbers because a thing may proceed in a numerical order, as in a man first the blood is generated, then the heart as the second thing, and then the other things.[298]

Yaḥyâ:
He says that he mentions number here by way of supposition, just as he uses letters by way of supposition when he wants to make clear what he is investigating.

[Aristotle 7.5]

{249b27-250b7 Proportionalities in motion.}

Yaḥyâ [ad 249b29]:
Everything that is causing motion[299] has caused motion before, because everything that is in motion has been moving a certain distance before, and what has moved has on account of the fact that it has been moving <taken a certain time to move>.[300]

249b27 What causes motion always moves something and this occurs in something and towards something. By 'in something'

I mean that the motion occurs in a time, and by 'towards something' I mean that a certain amount of distance is covered.

Yaḥyâ:
When everything that is in motion has a mover, and is moving during a certain time along a certain distance, we must investigate the relation between the motion and each of these things. He says: Suppose that a certain force moves something over a certain distance in a certain time. Let us take half of the force, while the time, the moved object and the distance remain the same. Then let us take half the moved object, while the others remain the same. Then let us take half the distance, while the others remain the same. Then let us take half the time, while the others remain the same.

[Aristotle 8.1]

{251b10-28 Motion and time in the universe have no beginning.

Yaḥyâ:[301]
It is not necessary that an instant is the end of previous time. We do 816,14
not assume that time is not generated. When he says that it is not reasonable to think that time might not exist, we say to this: saying that there is no time is similar to saying that outside the heavens there is no plenum, nor void, nor any extension at all, nor a time.[302]

[Aristotle 8.8]

Then Aristotle[303] shows that time is not composed of indivisible parts 906,23
(time atoms), and he first gives two premisses. The first premiss is that what exists, and did not exist before, must have been generated.
Yaḥyâ objects to a part of this statement, namely insofar as the 907,1
acquisition of a form is concerned[304] and he submits the statement to a condition, saying that what exists through a generation must have been generated. This is drawing a conclusion on the basis of the statement itself.

The second premiss is that what is being generated has not yet reached the state of having been generated, i.e. the form is not yet acquired while it is being generated.

Then he says:[305]

If we suppose a certain time consists of [parts] A, B, C and D, each of them indivisible, and something is turning white during the time AB and has become white at the time C, then it follows that between B and C there must be another time, because the form of the white has been acquired at the time C, whereas before that, i.e. during the time AB, it was not [yet] acquired, because that was the time of its generation, and a form is not yet acquired during the time of its generation. Thus, in the time AB it did not exist; it can exist in the time CD only if it has been generated [in a time] between AB and CD. We may give the same argument for that time as we did for this time, and this continues *ad infinitum*.

He says:

If I say that the forms are acquired at a now which is the end of the time of their generation and alteration, then it does not follow that between the now [of their acquisition] and the time [of their generation] there is another time, because I do not consider one now as being contiguous to, and succeeding another now, but [I say that] there is a time between each pair of nows. Nor do I consider a now as being something separate in itself, so that it could be said that the acquiring of a form in a now is subsequent to its generation. For one now is not contiguous to another now, so that we could claim it to be subsequent [to the preceding now].

Those who contradict this consider each of the [partial] times which make up the time ABCD as existing separately in themselves, and not being contiguous to one another.

I say that the argument is based on the premiss that the transition from generation to form is a transition from what is not-form to form. A process of such a kind can only be a generation, whether or not [the now] in which the form is acquired is contiguous to another.

His saying 'What exists and did not exist before must have been generated, and while it is being generated it does not exist' [263b26] ... .[306]

Notes

The works referred in the notes are given with their usual bibliographical data, except for the following:

Aristotle, *aṭ-Ṭabî'a*: Arisṭûṭâlîs, *aṭ-Ṭabî'a. Tarjamat Isḥâq ibn Ḥunayn ma'a šurûḥ Ibn as-Samḥ wa Ibn 'Adî wa Mattâ ibn Yûnus wa Abî l-Faraj ibn aṭ-Ṭayyib*, ed. A. Badawî, 2 vols, Cairo 1964-65

Giannakis, *Philoponus*: Giannakis, E., *Philoponus in the Arabic tradition of Aristotle's Physics*, thesis University of Oxford, 1992

Ibn Bâjja, *Commentary on the Physics*: Ibn Bâjja, *Šurûḥât as-samâ' aṭ-ṭabî'î*, ed. M. Ziyâda, Beirut 1978

Ibn Rušd, *Long Commentary*: Ibn Rušd, *Long Commentary on Aristotle's Physics*, Latin translation in: Aristotle, *Aristotelis omnia quae extant Opera. Averrois Cordubensis in ea opera omnes, qui ad haec usque tempora pervenere, commentarii*, vol. IV, Venetiis 1562, reprinted Frankfurt am Main 1962

Ibn Rušd, *Short Commentary*: Ibn Rušd, *Kitâb as-samâ' aṭ-ṭabî'î (Epitome in Physicorum libros)*, ed. J. Puig, Madrid 1983. Spanish translation as *Averroes, Epitome de fisica* by J. Puig, Madrid 1987

Lettinck, *Aristotle's Physics*: Lettinck, P., *Aristotle's Physics and its Reception in the Arabic world*, Leiden 1993

Philoponus, *in Phys.*: Philoponus, *In Aristotelis physicorum octo libros commentaria*, ed. H. Vitelli, *CAG* XVI-XVII, Berlin 1887-88

Ross, *Aristotle's Physics*: Ross, W.D., *Aristotle's Physics*, Oxford 1936

Simplicius, *in Phys.*: Simplicius, *In Aristotelis physicorum libros commentaria*, ed. H. Diels, *CAG* IX-X, Berlin 1882-95. English translation of Book 6 by D. Konstan, London and Ithaca NY 1989

Themistius, *in Phys.*: Themistius, *In Aristotelis physicorum libros paraphrasis*, ed. H. Schenkl, *CAG* V2, Berlin 1900

1. The lines 490,14-491,5 correspond to Philoponus, *in Phys.* 787,6-17.

2. The lines 491,8-9 are corrupt in the MS. A sensible text arises with our additions and deletions.

3. The sense of the text requires that one should read *muḥarrik* (mover) instead of *mutaḥarrik* (moved) at 493,14.

4. In *Physics* 3.3 Aristotle has shown that motion occurs in what is moved, not in the mover. Here he shows that motion occurs in what is moved, not in the states (forms) of the moving subject before and after the motion (cf. 224b25). See also Philoponus *in Phys.* 852,8-9.

5. The lines 496,5-16 correspond to Philoponus, *in Phys.* 852,9-18.

6. The lines 496,19-23 correspond to Philoponus, *in Phys.* 788,5-9 and 852,22-853,2.

7. Perhaps 417b6 ff.

8. The Greek text of Philoponus has: 'the ways which lead to knowledge are related to *generation rather than alteration*.' The Arabic version, as it appears in the MS, is not correct.

9. Things have homonymous names when they have a name in common but the definition of being (*logos tês ousias*) corresponding to the name is different (Aristotle, *Cat*. 1a1).

10. If 'affection' is used homonymously for form and for motion to the form, then one might say that form is an affection and that the motion to form is an affection. The conclusion that form is motion does not follow from these premisses. In order to be able to draw this conclusion one needs the premiss that an affection is a motion, but this cannot be said if 'affection' is used homonymously.

11. Plato, *Phaedo*, 70D-71A and 102E-104C

12. See 224a29: *kat' allên de kinêsin heteron – wa-qad yakûnu ġayran min qibali ḥarakatin uk̲râ*. Badawî has at 499,5: *wa-qad yakûnu ġayru aw mit̲l* instead of the correct text in the MS: *wa-qad yakûnu ġayran wa-mat̲t̲ala*.

13. This example corresponds to Aristotle's example of the statement that someone is going to Europe when he is going to Athens (224b22).

14. Read *laysa bi-kawnin* instead of Badawî's *laysa yakûnu* at 501,11.

15. The Greek *kai aei kai pantôn* is incorrectly rendered by Isḥâq as *wa huwa dâ'im lâ maḥâla* (it occurs certainly always).

16. The lines 501,22-5 correspond to Philoponus, *in Phys.* 788,14-18.

17. The MS has *ilâ mit̲lihî* (to a similar one). This does not fit in the context.

18. The lines 504,18-22 correspond to Philoponus, *in Phys.* 788,20-3.

19. The lines 505,2-3 correspond to Philoponus, *in Phys.* 788,25-7. At 505,2 the second *al-abyaḍ* (Badawî) should be *lâ abyaḍ* (MS).

20. The lines 506,2-19 correspond to Philoponus, *in Phys.* 788,29-789,8.

21. The context requires that 'not' is added.

22. A note in the margin of the MS says: 'Conjunction is a positive statement and separation is a negative statement, as if one separates something from something else.'

23. The Arabic has *bil-fi'l* (actual).

24. The lines 508,9-509,5 correspond to Philoponus, *in Phys.* 789,10-19 and 853,9-24.

25. Aristotle, *Physics* 190b16 ff.

26. The Arabic MS has *al-'araḍ* (the accident).

27. Aristotle, *Physics* 201a10.

28. The Arabic has *lawnan* (colour) instead of *kawnan*.

29. In Aristotle's Greek text (225b5 ff.) eight of the ten categories are mentioned: the categories of 'having' and 'being in a position' are omitted. The Arabic translation of Isḥâq mentions seven categories: the category of 'when' is also omitted.

30. The lines 513,7-12 correspond to Philoponus, *in Phys.* 789,21-9.

31. The Arabic text has *wa-takûnu nisbat al-muḍâf baynahumâ mawjûda* (while the relation of the relative between them is existing). This does not fit well in the context. Our translation is a conjecture.

32. This has been discussed in Aristotle, *Physics* 3.3 202b7ff. See also Philoponus, *in Phys.* 372,17-374,26.

33. Badawî incorrectly reads *taṣaffuḥ* at 516,23; the MS has *taṣaḥḥuḥ* (becoming healthy).

34. The MS has: 'The other way is that the motion is some other subject.' This phrase from Isḥâq's version is the first half of an incorrect translation of the Greek text 225b21, where Aristotle says: 'The other way is that some other subject might change from one change to another form.' The commentary however gives a correct rendering of Aristotle's statement.

35. Badawî incorrectly reads *sababan* (cause) at 517,3; the MS has *šay'an* (thing).

36. The word 'form' occurs in the second half of the phrase of which the first half is quoted here; see n. 34.

37. A lacuna the size of three words occurs here in the MS.

38. The text here has: becoming healthy.

39. The Greek MSS have *hugieian* (health) and that is how it is translated by Isḥâq. Philoponus and Simplicius apparently read *agnoian*, which has been adopted by Ross, see Ross, *Aristotle's Physics* 622.

40. The lines 519,14-19 correspond to Philoponus, *in Phys.* 853,29-854,4.

41. Badawî incorrectly reads *tantaširu* at 519,21; the MS has *tuṯîru*.

42. The lines 520,5-10 correspond to Philoponus, *in Phys.* 790,9-11.

43. Here 'simple' or 'absolute' generation is used in contrast to generation of generation.

44. Badawî incorrectly reads *sukûnan* (rest) at 523,4; the MS has *mutakawwinan* (generated).

45. This is one of Philoponus' arguments against infinite time. See the Introduction pp. 11-12 for a further discussion.

46. Our conjecture is *'abiṯat* instead of Badawî's *ġabiyat* (be ignorant) at 524,19; the word is partly without diacritical points in the MS.

47. This quotation is not from Isḥâq's version of the Greek text. It may be a free rendering of 226a13 in a reading which differs from the one chosen by Ross. Cf. Ross, *Aristotle's Physics* 624.

48. The lines 526,3-9 correspond to Philoponus, *in Phys.* 790,15-23.

49. The lines 526,11-14 correspond to Philoponus, *in Phys.* 854,6-10.

50. The lines 526,17-23 correspond to Philoponus, *in Phys.* 790,25-791,2.

51. The context requires that *fî* (in) is omitted.

52. The translation of Isḥâq has 'knowledge' (*'ilm*) instead of 'learning' (*ta'allum*), and it is quoted that way here in the commentary. The text of the commentary itself has the correct *ta'allum*.

53. The lines 527,9-14 correspond to Philoponus, *in Phys.* 791,7-12.

54. Badawî reads *šay'* (something) at 528,14; the MS correctly has *mašy*.

55. The lines 532,6-17 correspond to Philoponus, *in Phys.* 854,20-31 and 791, 14-15.

56. The problem is whether such changes are also between contraries.

57. The lines 533,8-15 correspond to Philoponus, *in Phys.* 854,32-855,4 and 791,17-20.

58. The term *akinêton* (*ġayr mutaḥarrik*) may also be translated as immovable; this term would be appropriate for the first three cases enumerated below, but not for the last case.

59. This is a variant of the name Yaḥyâ.

60. The lines 539,2-5 correspond to Philoponus, *in Phys.*

61. According to the Aristotelian text the condition for a motion to be continuous in this context – the definition of 'between' – is that it is not interrupted in its path; an interruption in time is allowed. Philoponus gives another interpretation here, which is senseless in this context: interruptions in the path are allowed, but no interruptions in the time. Themistius gives the same interpretation, cf. Themistius, *in Phys.* 172,26-173,8.

62. The lines 541,5-11 correspond to Philoponus, *in Phys.* 792,2-9 and 855,9-11.

63. The lines 542,2-3 correspond to Philoponus, *in Phys.* 855,13-14.

64. The lines 543,6-11 and 543,21-4 correspond to Philoponus, *in Phys.* 792,11-15 and 855,16-18.

65. The corresponding Greek text of Philoponus has: 'one town succeeds another town.'

66. The corresponding Greek text of Philoponus has: 'villages and mountains' instead of 'rivers, villages and deserts'.

67. The interpretation of the modern commentators of 227a2-3 is: '[Between things which are in succession there must be nothing of the same kind], such as between successive lines there must be no other line or lines and between successive units there must be no other unit or units.' The Arabic translation of Ishâq has: ' ... as a line or lines succeed a line or a unit or units succeed a unit', which makes very little sense in this context. Apparently Philoponus' interpretation is in agreement with that of Isḥâq. Simplicius also has this interpretation (see Simplicius, *in Phys.* 876,27-877,2).

68. The passage 227a7-10 is out of place in the Greek MSS of the Aristotelian text. It belongs to the discussion of 'between'.

69. After *bil-fi'l* (545,16) Badawî omitted from the MS: *iṯnayni bil-quwwa kâna muttaṣilan wa-ḏâlika anna nihâyat al-muttaṣilayni l-muttaḥidayni wâḥidatun bil-fi'l.*

70. Badawî has noticed a corruption in the MS text, but with his correction no sensible phrase arises. We have followed Giannakis' conjecture *al-munfaṣila bihâ yu'addu*, instead of Badawî's *al-muttaṣila minhâ ba'da* at 548,19, cf. Giannakis, *Philoponus* 248.

71. The context requires that *illâ* (except) should be added before *min* in 549,2.

72. The lines 549,3-5 correspond to Philoponus, *in Phys.* 792,26-793,1 and 855,20.

73. The lines 549,8-10 correspond to Philoponus, *in Phys.* 793,3 and 855,21-3.

74. The lines 549,12-14 correspond to Philoponus, *in Phys.* 856,5-6.

75. The lines 551,8-14 correspond to Philoponus, *in Phys.* 855,25-856,3.

76. The Arabic version has *wa-lâ l-ḥaraka min ḥarakât*, which clearly is not correct, but which is corrected by the next words. The corresponding Greek of Philoponus has *oute kinêsis ek kinêmatôn*, without any further explanation of what *kinêmata* are. Aristotle uses the word in 232a9 and 241a4, where he shows that motion is not composed of *kinêmata* (*taḥrîkât*), i.e. of starts (translation of Ross).

77. The lines 551,14-552,1 correspond to Philoponus, *in Phys.* 793,7-12.

78. Apparently change[k] itself is also considered to be a genus (category).

79. MS: *naw' lil-fâ'ida* (a species of what is useful), a corruption of an Arabic version of *eidos hupolêpseôs* (Giannakis suggests *naw' lil-qâ'ida* ?, see Giannakis, *Philoponus* 252). The editor of the texts contained in the Leiden MS has added to this passage the following phrase regarding the translation of *epistêmê eidos men hupolêpseôs* (227b13): 'In the translation of ad-Dimašqî: "knowledge falls under what is useful (*al-fâ'ida*)"; this is better than the translation of al-Isḥâq: "knowledge falls under thought (*aẓ-ẓann*)".'

80. The lines 554,15-18 correspond to Philoponus, *in Phys.* 793,14-19. The Arabic version seems to be corrupt. Even with the emendations indicated here the sense of the text is quite different from the Greek text, which runs as follows: 'One should note that if the path (*to eph' hou*) is different, then the manner (*tropos*) of the motion is different, e.g. rectilinear motion and circular motion. If the path is the same, then the manner may still be different, e.g. walking and rolling.'

81. The lines 555,7-9 correspond to Philoponus, *in Phys.* 793,27-9.

82. See 224a34-225b1.

83. The Greek text and Isḥâq's version (556,4) have: 'such as the white which becomes black.'

84. The Greek text has 'Coriscus' in 227b33; the Arabic translator has replaced this by 'Socrates'.

85. The lines 557,13-20 correspond to Philoponus, *in Phys.* 856,11-15 and 794,2-5. However, the Greek text gives the example that Coriscus is walking and becoming black, which is Aristotle's example. Besides the fact that the Arabic text has 'Socrates' instead of 'Coriscus', it has 'learning' and 'becoming hot' instead of 'walking' and 'becoming black'. The use of a different example is explained in the next passage. Apparently the Greek text of Philoponus has not only been summarised, but also adapted for the Arabic version by some editor.

86. The text has *an-nâr* (the fire), but *al-bârid* (the cold) is more fitting in the context.

87. The phrase 'although the time may be one and continuous' is not in Aristotle. It refers to the fact that time always flows continuously and cannot be interrupted, whereas a motion may be interrupted by another motion or by a rest, and in such a case cannot be considered as one motion anymore.

88. Accepting Giannakis' conjecture, *fa-kayfa*.

89. The text has *az-zamân* (the time), but 'the substrate' (*al-mawḍû'*) is more fitting in the context.

90. Alexander has an interpretation of 228a12 which is reviewed by Simplicius (*in Phys.* 889,7-16) which is different from what Philoponus attributes to Alexander here: 'If the substrate (*hupokeimenon*) of the conditions (*hexeis*) is different, the conditions are different; for one condition belongs to one substrate. The substrate is indeed different, as it does not remain the same because of the flux (*rhusis*). The relation of condition to activity (*energeia*) is like the relation of substrate to condition. For one activity belongs to one condition, but one condition may have many activities. They are many (different) because of the time intervals which separates them.'

91. The reading of this phrase in the Greek text is doubtful (cf. Ross, *Aristotle's Physics* 631-2). The Arabic translator apparently read: *plên tosouton diapherei, hoti ei men duo* (sc. *hai energeiai*) *di' auto touto houtôs tôi arithmôi kai tas hekseis anangkê* (sc. *duo einai*). This is quite different from what Philoponus apparently read (cf. Philoponus, *in Phys.* 856,17-24). The commentary of Yaḥyâ is in agreement with this reading of Philoponus. It asserts just the opposite of what is said in the Arabic version of Aristotle's text.

92. The last two phrases are from Yaḥyâ, according to Giannakis (cf. Giannakis, *Philoponus* 260). They are separated from the previous phrase by a remark from Abû 'Alî, and not preceded by the name Yaḥyâ, so they could be from Abû 'Alî as well.

93. 'Not' should be added, cf. 560,9.

94. The Greek *alla* in 228a24 has been omitted in the Arabic translation of Isḥâq and in this quotation by Yaḥyâ.

95. The lines 565,17-22 correspond to Philoponus, *in Phys.* 794,15-20.

96. The lines 565,25-566,6 correspond to Philoponus, *in Phys.* 794,22-9. However, the explanation of the torch-race is different in the Greek text of Philoponus: there are two rows of men; in each row a torch is passed on from one man to another; the winning row is the one in which the torch arrives first at the last man. In the Arabic version the race is held between two men; probably two *rows* of men are meant.

97. An incorrect translation by Isḥâq of the Greek text: the subject of the phrase should be 'the motion' instead of 'their time'. Yaḥyâ's commentary assumes the correct version.

98. The lines 567,5-11 correspond to Philoponus, *in Phys.* 856,26-857,4.

99. Arabic: *al-ajrâm* (bodies), i.e. the celestial bodies (*al-ajrâm al-mustadîra*, lit.: the circular bodies). It is clear from the context that by *al-ajrâm* the celestial bodies are meant.

100. i.e. the heavens rotate uniformly because they are endowed with intellect, or rather, intellects.

101. The Greek *kata genos* (in genus) here does not occur in the Arabic version, but the commentary does take it into account.

102. The lines 568,14-19 correspond to Philoponus, *in Phys.* 857,6-15 and 794,31-795,2.

103. The lines 571,16-21 correspond to Philoponus, *in Phys.* 795,11-18.

104. These words are a conjecture. The phrase in the text seems to be incomplete.

105. The Greek has: *en tôi pou*; the usual Arabic translation is: *fî l-ayn* (in the 'where').

106. The lines 572,11-15 correspond to Philoponus, *in Phys.* 795,7-9.

107. After *wa-immâ aṯqal* (572,21) Badawî omitted from the MS: *al-awlâ bi-hâḏâ an yartaqiya ilâ šay' wâḥid*.

108. The lines 572,23-573,4 correspond to Philoponus, *in Phys.* 857,17-20.

109. The lines 573,11-19 correspond to Philoponus, *in Phys.* 857,22-5 and 795,20-6.

110. The MS has *aḍ-ḍiddayn* instead of Badawî's *aḍ-ḍidd* at 579,18.

111. The lines 581,1-5 correspond to Philoponus, *in Phys.* 795,32-796,5 and 857,32-858,4.

112. The lines 581,12-14 correspond to Philoponus, *in Phys.* 858,4-7.

113. The lines 581,20-582,2 correspond to Philoponus, *in Phys.* 858,16-23.

114. The lines 582,13-19 correspond to Philoponus, *in Phys.* 858,16-23.

115. The lines 586,2-15 correspond to Philoponus, *in Phys.* 796,21-5 and 858, 25-8.

116. cf. *Cat.* 12a26 ff., especially 13a33 ff.

117. The lines 586,18-22 correspond to Philoponus, *in Phys.* 858,30-4.

118. Apparently Yaḥyâ interprets 'these two' (*hautai*) in 229b32 as referring to the contrary motions mentioned in the preceding phrase, not to the rests, which is the usual and correct interpretation.

119. i.e. whether the absence of change is a state of rest.

120. Before *illâ innahû* (589,20) Badawî omitted from the MS: *Arisṭû*.

121. The quotation omits a third alternative which occurs in the Greek text (i.e. that the absence of change in a thing has as its contrary the absence of change in the thing's not-being). This alternative is absent in the Arabic version of Isḥâq. The commentary of Yaḥyâ however does mention this alternative.

122. The line 590,7 corresponds to Philoponus, *in Phys.* 797,12.

123. This comment is not preceded by the name Yaḥyâ; it is clear, however, that it must be by him. It corresponds to Philoponus, *in Phys.* 797,14.

124. The lines 594,1-5 correspond to Philoponus, *in Phys.* 797,19-26.

125. After *laysa minhumâ* (594,6) Badawî omitted a line from the MS: *šay' k̲ârij 'an aṭ-ṭabî'a wa-huwa yaḥullu l-šakk bi-anna kull hâd̲ihî l-ḥarakât fîhâ*.

126. i.e. the critical day in the normal course of an illness.

127. The lines 594,18-22 correspond to Philoponus, *in Phys.* 859,10-14.

128. The lines 594,22-4 correspond to Philoponus, *in Phys.* 859,16-18.

129. Alteration is the technical term for change of quality. In this passage illness and health are not considered as attributes of a body (change between attributes *would be* alteration, one of the three kinds of change[k]) but as privation and form. Thus change between them is generation and corruption, not *kinêsis*. Cf. *Phys.* 5.2, 226a27.

130. The line 596,1 corresponds to Philoponus, *in Phys.* 797,29-30.

131. The lines 596,3-4 correspond to Philoponus, *in Phys.* 859,16-18.

132. The lines 596,6-8 correspond to Philoponus, *in Phys.* 859,20-2.

133. The lines 596,11-14 correspond to Philoponus, *in Phys.* 859,23-7.

134. This is not a literal quotation from Aristotle's text. It may refer to 230a29 ff.

135. Read *sukûnuhû* instead of *ḥarakatuhû* (its motion).

136. Omitting *sukûnuhâ* in 596,19.

137. In some manuscripts of the Greek text this phrase is added in 230b12 between *hautai* and *pheretai*. It occurs in Isḥâq's translation – and in the Latin version of Ibn Rušd's *Long Commentary*. Ibn Rušd used Isḥâq's version.

138. The lines 596,24-597,8 correspond to Philoponus, *in Phys.* 798,4-12.

139. The lines 599,5-19 correspond to Philoponus, *in Phys.* 798,14-799,2 and 859,29-860,5.

140. The Greek text has: *to thatton kineisthai* (e.g. 798,15); this is rendered in Arabic as: *sur'a hâd̲ihî l-ḥaraka*, lit.: the velocity of this motion.

141. The lines 600,11-16 correspond to Philoponus, *in Phys.* 799,4-9.

142. The Greek text gives a clearer version of the phrase between dashes: 'For it is possible that the motion of a thing to its place is not a coming to a halt, i.e. an accelerated motion, right from the start.'

143. The lines 601,9-12 correspond to Philoponus, *in Phys.* 860,11-15.

144. The lines 601,15-602,2 correspond to Philoponus, *in Phys.* 860,20-7.

145. The lines 602,5-6 correspond to Philoponus, *in Phys.* 860,28-31.

146. The lines 602,9-10 correspond to Philoponus, *in Phys.* 860,33-5. 791,22-7.

147. For discussion of Aristotle's argument, see R. Sorabji, *Time, Creation and the Continuum*, London and Ithaca NY 1983, 365-9.

148. The lines 606,6-10 correspond to Philoponus, *in Phys.* 861,3-9.

149. The Greek term is *kinêmata* and the Arabic one is *taḥrîkât*; they occur in Aristotle's text at 232a9 and 241a4. These terms refer to that which has the same relation to a motion as a point has to a line and an instant to a time interval. It may rendered in English as 'start', 'having-movedness', 'instantaneous leap' or '(instantaneous) jerk'. It corresponds to the successive instantaneous states of the changing object. The idea is that of cinematographic disappearance from one place and reappearance in another.

150. The lines 606,12-15 correspond to Philoponus, *in Phys.* 800,3-6.

151. The lines 607,17-18 correspond to Philoponus, *in Phys.* 800,11-12 and 861,11-12.

152. The lines 608,3-6 correspond to Philoponus, *in Phys.* 800,18-20 and 861,15-17. The Greek text gives as an example of things which are different in definition but not in place: honey which is sweet and yellow. The Arabic version has instead the white, learned man.

153. The lines 608,7-8 correspond to Philoponus, *in Phys.* 800,11-12 and 861, 11-12.

154. The lines 609,13-19 correspond to Philoponus, *in Phys.* 861,19-23.

155. On jerks, see Richard Sorabji, op. cit., 365-71 with 17-21, 52-6 and 345-8.

156. sc. an instant.

157. sc. a jerk, or a momentary state of the changing object.

158. cf. Aristotle, *Phys.* 219a10 ff. and 220a4 ff.

159. The following proof establishes only the converse, that indivisible spaces would need to be traversed by indivisible jerks (disappearance from one space followed by reappearance in another space). It is not true that such cinematographic motion would, in turn, imply indivisible spaces.

160. Cinematographic disappearance and reappearance elsewhere implies that something *has* moved to its next postion, without its ever being true that it *is* moving to its next postion. This is alleged to be impossible.

161. Aristotle uses the example of someone walking to Thebes (231b30). In the Arabic version of Philoponus' commentary this is changed to someone walking from Baghdad to Basra. Something similar may be found in the commentary on Book 3, ch. 3, see the Introduction, p. 7.

162. The lines 614,16-20 correspond to Philoponus, *in Phys.* 863,2-7 and 801, 19-22.

163. For comment, see Richard Sorabji, op. cit., 365-9.

164. The lines 619,16-620,1 correspond to Philoponus, *in Phys.* 863,9-15.

165. These three things are not the same as those occurring in Aristotle's text 232a25-7. The first one is not mentioned by Aristotle. What Aristotle mentions as the third one is not among the three mentioned here, but is mentioned later, after the phrase: 'He also says ... ' The translation of Isḥâq omits Aristotle's second case, i.e. *en tôi elattoni ison*.

166. This paragraph makes sense only if we suppose that it is concerned with the example given directly above.

167. The lines 624,8-11 correspond to Philoponus, *in Phys.* 802,5-8.

168. The lines 625,17-18 correspond to Philoponus, *in Phys.* 802,22 and 863, 28-9.

169. The lines 627,12-628,4 correspond to Philoponus, *in Phys.* 863,31-4 and 802,25-30.

170. The lines 628,5-10 correspond to Philoponus, *in Phys.* 864,2-9.

171. The lines 628,22-629,15 correspond to Philoponus, *in Phys.* 802,32-803,12.

172. i.e. a multiple of C equals A.

173. The MS has: 'two-thirds' instead of 'three-quarters'.

174. The lines 632,22-633,3 correspond to Philoponus, *in Phys.* 803,14-19.

175. This last example is given in the Greek text of Philoponus. According to it, if one takes two cubits (as a part from the whole magnitude of seven cubits), then four times this measure will be greater than the whole and three times will be less. The Arabic version is quite different and shows a misunderstanding of what is meant by *elleipsei* and *huperbalei*.

176. The name of the commentator is not mentioned.

177. The name of the commentator is not mentioned.

178. This is a comment is on 233b15 ff., which belongs to the next section.

179. One should add from the MS: *yanqasimu* (would be divisible) after *yakûnu mâ lâ yanqasimu* in the edition of Badawî 636,18.

180. The lines 640,13-14 correspond to Philoponus, *in Phys.* 803,27-30.

181. The lines 643,1-2 correspond to Philoponus, *in Phys.* 804,7-8 and 864,20-2. The phrase is not finished in the Arabic version. One may complete it from the Greek text as follows: ... and every time is divisible, one may divide the now at D.

182. This unfinished phrase is misplaced; it may be the beginning of the commentary on the next lemma.

183. This comment and the following four are not preceded by the name Yaḥyâ. See also note 186.

184. The lines 643,11-14 correspond to Philoponus, *in Phys.* 803,30-804,2 and 864,13-18.

185. The translation of Isḥâq and the commentary of Yaḥyâ show an interpretation of the Aristotelian text which differs from that of other commentators. Simplicius (*in Phys.* 955,20 ff.) and Themistius (*in Phys.* 189,25ff.) have the following interpretation, which they share with the modern commentators: Nothing of the future belongs to this side of the now (i.e. the side of the past) and nothing of the past belongs to that side of it (i.e. the side of the future). Ibn Rušd (*Long Commentary* 262B) follows Isḥâq and Philoponus.

186. The lines 644,1-3 correspond to Philoponus, *in Phys.* 804,7-11 and 864,20-5. This is an indication that this comment, which is not preceded by 'Yaḥyâ', is a paraphrase of one of Philoponus' comments, and this will also be true for the four preceding comments.

187. The MS has *qabla* instead of Badawî's *mitl* at 646,9.

188. The MS has *al-mašy* instead of Badawî's *aš-šay'* (the thing) at 647,3.

189. After *annahû sâkin* (647,9) Badawî omitted one line from the MS: *li-annahû ġayr mâšin wa-lâ yuqâlu ayḍan fîmâ min ša'nihî ḥarakat al-istidâra innahû sâkin*.

190. Badawî omitted 'not' from the MS at 649,9.

191. Aristotle mentions *pêxis* (coagulation or freezing) as an instantaneous change in *Phys.* 253b25 and *De Sensu* 6, 447a1-3.

192. See the Introduction, pp. 13-15 for a further discussion of this problem of instantaneous change. See also Richard Sorabji, op. cit., 410-11.

193. The Arabic text has 'motion' (*ḥaraka*) instead of 'what is in motion' (*mutaḥarrik*).

194. The lines 654,18-656,1 correspond to Philoponus, *in Phys.* 805,19-806,18.

195. The Arabic text of the preceding two phrases is not logically consistent. By emending a few letters designating parts of the motion GH one gets a sensible text which corresponds to the Greek text of Philoponus. This runs as follows (806,3ff.): 'If DE is not the motion of AC, then suppose it is GH. Subtract from GH the motion of the two parts AB and BC of AC. Then GH, which is thus divided into parts, either exhausts them, or falls short of them, or surpasses them. If it surpasses them, such as when GI is the motion of AB and IK the motion of BC, then KH remains, and this must be motion of something.'

196. This problem is discussed in the Introduction, pp. 15-16.

197. The MS has *lâ annahû* instead of Badawî's *illâ annahû* at 656,22.

198. Badawî omitted *ḥattâ* from the MS at 656,22.

199. i.e. parts which have the same degree of whiteness as the whole.

200. A *raṭl* is a weight (about 0.45 kg in Egypt), which is divided into 12 *wiqiyya*.

201. MS: *fa-in lam ya'tabir hâḏâ l-mutašakkik* instead of Badawî: *fa-in lam ya'tari hâḏâ l-maslak* (658,13).

202. Simplicius *in Phys*. 979,7.

203. The lines 659,4-5 correspond to Philoponus, *in Phys*. 865,8-9.

204. The line 659,8 corresponds to Philoponus, *in Phys*. 865,19-20.

205. The lines 662,8-10 correspond to Philoponus, *in Phys*. 865,25-7.

206. cf. Ross, *Aristotle's Physics* 647

207. The line 662,12 corresponds to Philoponus, *in Phys*. 865,27-8.

208. This phrase seems to be a gloss; the previous words 'and of parts of FH' make no sense.

209. The lines 664,9-14 correspond to Philoponus, *in Phys*. 806,23-7.

210. The lines 664,23-665,2 correspond to Philoponus, *in Phys*. 806,31-807,3.

211. The lines 665,10-13 correspond to Philoponus, *in Phys*. 866,2-6 and 807, 15-17.

212. This is the first of several passages relevant to the first and last instants of rest and change. For discussion, see Richard Sorabji, op. cit., ch. 26.

213. The lines 667,8-12 correspond to Philoponus, *in Phys*. 866,8-13 and 807, 19-21.

214. The lines 668,16-17 correspond to Philoponus, *in Phys*. 866,18-19.

215. The lines 669,15-670,3 correspond to Philoponus, *in Phys*. 807,30-808,6 and 866,26-31.

216. The lines 670,4-6 correspond to Philoponus, *in Phys*. 808,6-10.

217. The lines 670,9-11 correspond to Philoponus, *in Phys*. 866,33-5.

218. The lines 670,14-21 correspond to Philoponus, *in Phys*. 866,35-867,7.

219. The lines 672,16-20 correspond to Philoponus, *in Phys*. 867,12-14 and 808,12-14.

220. The word translated as 'completion of a change' is without diacritical points in the MS; Badawî reads it as *bu'd* (remoteness, distance) at 672,19, which gives no sensible phrase. We conjecture *taġayyur*.

221. The lines 673,4-8 correspond to Philoponus, *in Phys*. 808,16-19.

222. This remark does not make sense in the context. Also, it is not in agreement with the corresponding Greek text of Philoponus, who says that instead of *meteballen* (235b37) one should read *metaballei*. The rest of the comment does not clarify this question; the Greek text as well as its Arabic version is just an account of 235b37-236a2.

223. The lines 674,1-2 correspond to Philoponus, *in Phys*. 808,21-2.

224. The MS has *wa-annahû lam yakun min ba'dihî* instead of Badawî's *wa-in lam yakun min hâḏihî* at 676,3.

225. The lines 677,4-6 correspond to Philoponus, *in Phys*. 809,2-4.

226. The corresponding Greek text is: 'As he has taken AD as indivisible, he thinks that one could as well say (*adiaphoron eipein*) "in D" [instead of in A].' The Arabic version is incorrect: *adiaphoron* is rendered as *ġayr munqasim* (indivisible).

227. The lines 677,9-11 correspond to Philoponus, *in Phys*. 809,6-7 and 867,33-4.

228. The lines 677,14-17 correspond to Philoponus, *in Phys*. 809,9-13.

229. The lines 678,9-15 correspond to Philoponus, *in Phys*. 868,6-18.

230. The objection brought forward here apparently refers to the case of local motion. Then, indeed, one cannot say that a part of the moving object moves during a part of the time, because the object moves as a whole, all its parts simultaneously being in motion. In fact, the proposition discussed here does not apply at all in this case, and there can be no 'solution' of the objection. The following phrases containing a 'solution' make no sense; it is difficult to imagine which solution Philoponus may have had in mind.

231. The lines 691,13-17 correspond to Philoponus, *in Phys*. 869,22-5.

232. The lines 692,10-22 correspond to Philoponus, *in Phys*. 812,5-15.

233. The line 696, note 1 corresponds to Philoponus, *in Phys*. 870,30-1.

234. in 233a13 ff.

235. The MS has *laysa yumkin* (not possible); we follow Giannakis' conjecture *laysa yunkar* (see Giannakis, *Philoponus* 347).

236. The MS has *al-kalâm* (the statement). Giannakis' conjecture is *as-sukûn* (see Giannakis, *Philoponus* 348).

237. The lines 699,3-6 correspond to Philoponus, *in Phys*. 812,25-7.

238. Apparently the Aristotelian text 237b33 ff. is interpreted in such a way that the motion under discussion is not a local motion, but a qualitative change of a body. Then 'stretch' in the comment refers to a part of the changing body, not of the path along which a local motion would occur. The Greek text allows for both interpretations; the Arabic translation has explicitly taken the interpretation of a qualitative change. In the next comment Philoponus says that both interpretations are possible.

239. The lines 699,14-16 correspond to Philoponus, *in Phys*. 812,29-813,2.

240. The MS has *al-fi'l* (the effect); the Greek has *megethos*, so we conjecture *al-'iẓam*.

241. The lines 699,19-20 correspond to Philoponus, *in Phys*. 813,5-7.

242. The lines 703,13-14 correspond to Philoponus, *in Phys*. 817,18-19.

243. The lines 703,16-20 correspond to Philoponus, *in Phys*. 871,21-31.

244. The word is unclear in the MS; we follow Giannakis' conjecture *makâniyya* (see Giannakis, *Philoponus* 350).

245. The lines 711, note 1 correspond to Philoponus, *in Phys*. 817,6.

246. The lines 715, note 2 correspond to Philoponus, *in Phys*. 817,20-7.

247. The text makes sense if we replace *stadion* by 'row of bodies'. Aristotle uses *ongkoi* for the rows of bodies AA and BB. It is not clear why Philoponus says that there are *stadia* AA and BB. The Greek text of Philoponus supplies the information that AA is not moving and BB is moving.

248. This comment (lines 722,8-12) is not preceded by a name; it must be Yaḥyâ's comment, for it corresponds well to a passage from the Greek text of Philoponus, cf. *in Phys*. 817,8-10 and 16-18.

249. The situation is illustrated by the following figure:

```
        AAAAAA              AAAAAA

BBBBBB→                     BBBBBB→

            ←CCCCCC     ←CCCCCC
```

250. The lines 729,7-11 correspond to Philoponus, *in Phys*. 819,2-5.

251. The lines 730, note 1 correspond to Philoponus, *in Phys*. 820,17-19.

252. It is well-known that there are two versions, Alpha and Beta, of Book 7, Chs. 1-3 (cf. Ross, *Aristotle's Physics*, 11ff.). The translation of Isḥâq is a translation of version Beta. Our references to Aristotle's text are to this version Beta, if not indicated otherwise. As has been said in the introduction, quotations of Aristotle in Yaḥyâ's commentary are from Isḥâq's translation, so here they are in version

Beta. However, also quotations from version Alpha occur. In most cases the phrases quoted from one version have a corresponding phrase in the other version, but in a few cases a phrase occurs in Alpha only. The Greek fragments of Philoponus are also comments on passages from Alpha and from Beta.

When for a quotation of Aristotle no version is mentioned, it is from Isḥâq's version (Beta), but it also agrees with version Alpha; when a version is mentioned, the quotation is from that version only.

253. This comment (p. 734, note 1) is not preceded by a name; it must be Yaḥyâ's comment, for it corresponds well to a passage from the Greek text of Philoponus, sc. *in Phys*. 874,5-6 and 822,3-4.

254. From the proposition 'If a then b' (or: every a is b) one may derive the converted proposition 'If ~b then ~a (or: every not-b is not-a) (~a is the negation of a).

255. *Laysa* (not) should be added in the text.

256. The MS clearly has *šam'* (wax).

257. The text has 'fire' instead of our conjecture 'water'.

258. Version Alpha lacks the word 'else'. The presence of this word in Beta is used by Robert Wardy to suggest that Book 7 is not an independent treatise, but part of the support for Book 8, Chapter 5. See Robert Wardy, *The Chain of Change*, Cambridge 1990.

259. The lines 744,21-2 correspond to Philoponus, *in Phys*. 874,9.

260. The lines 744,25-745,2 correspond to Philoponus, *in Phys*. 874,17-20.

261. i.e. natural motion. In fact, this is written in the MS between the lines.

262. cf. Philoponus *Corollary on Void*, *in Phys*. 679,33-680,23.

263. Badawî omitted *ġayr* from the MS at 755,1.

264. This explanation of Aristotle's text is just the opposite of the interpretation of the modern commentators.

265. In *Cat*. 8b25-9b9 four kinds of qualities are distinguished: shape, state or condition, capacity/incapacity and affective quality.

266. The MS has *fa-l-abṣâr lâ* ... instead of Badawî's *fa-l-a'ḍâ'* (the members) at 756,17; the sense of the phrase is not clear.

267. The text has *al-anwâ'* (the kinds); we conjecture *ar-rawâ'iḥ*.

268. The lines 757,18-20 correspond to Philoponus, *in Phys*. 876,10-11.

269. The phrase seems to be unfinished in the Arabic text.

270. These two phrases (lines 767,16-18) correspond to Philoponus, *in Phys*. 876,25-6 and 29.

271. A laudable pain is e.g. shame, according to a comment of Abû Bišr at 768,26.

272. i.e. heat etc. According to the comment of Abû Bišr at 768,24-5 heat is gathered inside the body when a desire is forbidden, and it spreads outwards when one favours it.

273. The lines 769,25-770,2 correspond to Philoponus, *in Phys*. 877,2-5.

274. The lines 770,3-8 correspond to Philoponus, *in Phys*. 877,8-17.

275. The lines 770,13-22 correspond to Philoponus, *in Phys*. 877,19-27.

276. See n. 265 above.

277. The context requires this addition to the MS.

278. i.e. the species name must mean the same thing as applied to each change.

279. This phrase as it appears in the MS contains signs indicating that certain words should be omitted, however not in a consistent way. We give here the most sensible text which can be extracted from the MS; it differs from the reading of Badawî.

280. Badawî reads (twice in 783,64) *taḥdabu* (be convex); this should be *tuḥdaṯu* (is produced). The word is without diacritical points in the MS.

281. Apparently the angle formed by a straight line and a curved line is not considered to be equal to the angle between that straight line and the tangent of the curved line, but (infinitesimally) larger or smaller. This is what is implied by proposition II,16 of Euclid's *Elements*, as has been remarked by Giannakis (cf. Giannakis, *Philoponus* 377). See T.L. Heath, *The Thirteen Books of Euclid's Elements*, Cambridge 1956 (2nd edition), vol. II, 37-43.

Especially interesting is the view of Johannes Campanus (thirteenth century), mentioned by Heath: If a line turns on the intersection of a circle and its diameter, the angle between the diameter and that line is always smaller than the angle between the diameter and the semi-circle, as long as the line intersects the circle at a second point, and it is larger than that angle as soon as it has become the tangent; thus, the angle between diameter and moving line turns from being smaller than a certain angle to larger than it, without ever becoming equal to it.

282. The lines 784,4-6 correspond to Philoponus, *in Phys*. 877,34-5.

283. The lines 786,3-5 correspond to Philoponus, *in Phys*. 878,8-12.

284. *to diêkhes* in Philoponus' Greek text 878,12. This refers to the elasticity of the air or its capacity to vibrate. The Arabic version incorrectly has *ṣafâ'* (clearness).

285. The lines 786,8-24 correspond to Philoponus, *in Phys*. 878,22-30.

286. The lines 787,1-3 correspond to Philoponus, *in Phys*. 878,32-879,2.

287. The lines 787,6-10 correspond to Philoponus, *in Phys*. 879,4-8.

288. The Arabic has *al-qqns*, a transliteration of *kuknos* (swan). Giannakis suggests this could be a corruption of a transliteration of the Greek *kunos*, since it occurs in the Greek text. The Arabic version of Yaḥyâ gives the whiteness of a horse and a dog at 785,3,22 and 23; see Giannakis, *Philoponus* 382.

289. The Arabic has *bil-'adad* (in number), whereas the Greek correctly has *kat' eidos*.

290. The lines 797,21-798,15 correspond to Philoponus, *in Phys*. 879,25-880,15.

291. The Greek text of Philoponus in this phrase has 'two garments' (*duo esthêtes*) instead of 'two individuals' (*šaḵṣâni*).

292. The Greek text of Philoponus has 'becoming healthy' and 'becoming white' (880,1-3) instead of 'becoming white' and 'becoming black'.

293. The part of the phrase between the last two commas has been omitted by Badawî at 799,17.

294. The Arabic text has diminution (*naqṣ*), but this would contradict what is said a few phrases later.

295. The Arabic text has *muttaṣil* (continuous); this should be *mutaḍâmma*.

296. This is the translation of the Arabic text, in which 'general' ('*âmma*) and 'particular' (*ḵâṣṣa*) belong to 'processes of generation'. This gives no sensible meaning. One should take the words '*âmm* and *ḵâṣṣ* in the same way as in Aristotle's text 249b25. Then the phrase runs as follows: 'He does this because there is no name which might be applied to all processes of generation [to indicate that they are different]; there is not such a name for both relations [of being different] commonly [such as dissimilarity or inequality], nor are there names for each specific relation [such as more and less or greater and smaller], so that the processes of generation could easily be divided and it could be made clear which ones are equal and which ones are not equal.'

297. The text has *al-ḥayawân* (animal); the context requires *al-insân*.

298. It is not clear whether this account of Alexander's commentary belongs to Philoponus' commentary, or is added by the compiler of the commentaries in the Leiden MS. Simplicius does not mention such a commentary by Alexander.

299. The Arabic text has *mutaḥarrik* (what is in motion); we emend it into *muḥarrik* (what causes motion) on account of the Aristotelian text.

300. The phrase is incomplete in the Arabic text; we conjecture these words on the basis of the Aristotelian text.

301. This commentary is an interruption of Abû l-Faraj's commentary on 251b10-28. It is preceded by the words: Yaḥyâ objects to this, saying ...

This commentary of Abû l-Faraj is an account of Aristotle's passage 251b10-28, in which he says that motion must be eternal because time is eternal, and time is a number of motion. Time is eternal because any instant is the end of preceding time and the beginning of following time, so that there can never be an instant which might serve as the beginning of time only. It is against this statement that Yaḥyâ raises an objection.

There is no Greek fragment of Philoponus' commentary corresponding to this passage. See the Introduction p. 17 for a further discussion.

302. Thus, if the latter claim is reasonable, which is the case according Aristotle, the former is reasonable too, and then it is not impossible that time is generated.

303. Like the preceding comment, the commentary of Yaḥyâ appears as an interruption of Abû l-Faraj's commentary. We give the whole passage in which this interruption occurs.

304. An individual form of whiteness can come into existence without going through a process of being generated, as Philoponus argues in his *Against Proclus on the Eternity of the World*, 340; 347; 365,3 and *Against Aristotle on the Eternity of the World* ap. Simplicium *in Phys.* 1141. Cf. Philoponus *in Phys.* 54,24-5; 55,12-134; 191,9-33.

305. The lines 907,6-19 correspond to Philoponus, *in Phys.* 845,34-846,19.

306. The phrase is corrupt and incomplete in the text. We have followed the Aristotelian text.

English-Arabic-Greek Glossary

Not all occurrences are mentioned, only those in key passages and/or the first occurrence.

The Greek glossary gives the word used by Aristotle in the passage on which Yaḥyâ's text is a comment (between brackets if the word does not occur at that passage, but somewhere else in the Aristotelian corpus). Philoponus uses the same word in his commentary, as far as it is extant.

The Arabic glossary gives the word used by Yaḥyâ in his commentary. 'I:' indicates the word used by Isḥâq in his translation of the passage which is commented upon, if it is different from the word used by Yaḥyâ.

The references to Book 7, chs 1-3 are to version beta, which is the one translated by Isḥâq.

accidental – *bil-'araḍ* 491,2 – *kata sumbebêkos* 224a21
action – *fi'l* 513,13 – *poiein* 225b7
affection – *aṯar* 496,22 I: *ḥadaṯ* – *pathos* 224b11
  ḥadaṯ 498,10 – *pathos* 224b14
  infi'âl 561,12 – *pathos* 228a8
  aṯar 796,22 – *pathos* 249b5
alteration – *istiḥâla* 529,15 766,12 – *alloiôsis* 226a26; 245b20
amber – *kahrabâ'* 755,24 (– *êlektron* 388b19; 736a5)
ancestors – *âbâ'* 523,12
apart – *furâdâ* 538,18 – *khôris* 226b18

between – *mâ bayna* 537,9 – *metaxu* 226b19

capacity – *qûwa* 766,15 (– *dunamis* 9a15)
carrying – *ḥaml* 753,19 – *okhêsis* 243a25
categories – *maqûlât* 490,9; 512,8 – *katêgoriai* 225b5
change – *taġayyur* 490,10 – *metabolê* 224a21
comparable – *mutaḍâmm* 783,9; 785,6 – *sumblêtos* 248a11
condition – *ḥâl* 766,14; 767,6 I: *hay'a* – *hexis* 245b22
contiguous – *šâfi'* 537,10 – *ekhomenos* 226b20
continuous – *muttaṣil* 537,10 – *sunekhês* 226b20
contradictory, contradiction – *mutanâqiḍ* 501,1 – *antiphasis* 224b29
contrary – *ḍidd* 501,14; 579,11; 586,8 – *enantios* 224b29; 229a7; 229b26
  muḍâdd 580,17; 586,6 – *enantios* 229b1 229b23
  mutaḍâdd 501,1; 579,13; 586,12 – *enantios* 225a12; 229b3; 229b32
corruption – *fasâd* 504,12 – *phthora* 225a18

diminution – *naqṣ* 531,4 – *phthisis* 226a31
disposition – *malaka* 766,14; 767,7 I: *hay'a* – *hexis* 245b22
divisible – *munqasim* 650,12; 654,9 – *diaretos* 234b10; 234b21

essential – *biḏ-ḏât* 491,2 – *kath'auto* 224a28

finite – *mutanâhin* 629,2 – *peperasmenos* 233a23
form – *ṣûra* 496,20 – *eidos* 224b11
ḵilqa 766,12 I: *ṣûra* 245b7 – *morphê* 245b22

generation – *kawn* 490,11; 504,9; 506,2; 522,17 I: *takawwun* – *genesis* 225a13; 225a26
takawwun 503,11
growth – *numûw* 531,4 – *auxêsis* 226a31

halting – *tawaqquf, taskîn* 599,7 – *histasthai* 230b22
having – *lahû* 512,9 (– *ekhein* 1b27).
homonymous – *muštarik* 498,12
muttafiq 784,21, *muttafiq ismuhû* 785,1 I: *muttafiq fî l-ism* – *homônumos* 248b7

indivisible – *ġayr munqasim* 640,13 – *adiaretos* 233b34
infinite – *lâ nihâya (lahû)* 523,12; 628,5; 629,1; 698,13; 730,n – *apeiros* 238a33
bi-lâ nihâya 628,8; 698,13 – *apeiros* 237b24
ġayr mutanâhin 628,19; 698,14 – *apeiros* 233a17; 238a1
I: *sarmad* – *apeiros* 241a26
I: *bi-ġayr nihâya* – *apeiros* 241b13
irregular – *muḵtalif* 567,12 – *anômalos* 228b16

knowledge – *'ilm* 552,24 – *epistêmê* 227b13
'ilm 769,17 I: *ma'rifa* – *epistêmê* 247a29

magnet – *ḥajar al-maġnâṭîs* 755,24
ḥajar 756,5 (– *lithos* 267a2)
motion – *ḥaraka* 510,11 – *kinêsis* 225b2
motion: local motion – *naqla* 531,13 – *phora* 226a33
motionless – *ġayr mutaḥarrik* 534,13 – *akinêtos* 226b10
moved – *mutaḥarrik* 493,9 – *kinoun* 224a34
mover – *muḥarrik* 493,9 – *kinoumenon* 224a34

now – *ân* 640,13 – *nun* 243b33

partial – *bil-juz'* 491,2 – *kata merê* 224a25
passion – *infi'âl* 513,13 – *paskhein* 225b7
perishing – *fasâd* 504,12 – *phthora* 225a18
position – *mawḍû'* – (*keisthai* 1b27)
pulling – *jaḏb* 753,26 – *helxis* 243a25
pushing – *daf'* 753,24 – *ôsis* 243a24
pushing on – *sawq* 754,1 – *epôsis* 243a26
pushing off – *zajj* 754,2 – *apôsis* 243a27

quality – *kam* 529,7 – *poson* 225b7
quantity – *kayf* 529,7 I: *kayfiyya* 225b6 – *poiotês* 225b6, *poion* 226a24

relation – *muḍâf* 513,7; 767,17 – *pros ti* 225b6; 246b4
rest – *sukûn* 586,3 – *êremia* 229b24
rotation – *ḥaraka dawriyya* 730,n I: *naqla yakûnu dawran* – *kuklôi phora* 241b20

shape – *šakl* 766,12 – *skhêma* 245b22
stadium – *mîdân* 715,n – *stadion* 239b35
state – *hay'a* 561,11; 767,6; 768,14; 769,14 – *hexis* 228a8; 245b22; 246a29
substance – *jawhar* 513,6 – *ousia* 225b5
succession: in succession – *mâ yatlû* 537,9 – *ephexês* 226b19
synonymous – *mutawâṭi'* 784,21 – *sunônumos* 248b7 in the version quoted by Simplicius, *in Phys.* 1086,20

throwing – *ramy* 754,5 – *rhipsis* 244a21
together – *ma'an* 538,4 – *hama* 226b18

touching – *mumâss* 537,9 I: *liqâ'*
mumâss – *haptomenos* 226b19
mumâss 549,3 I: *mulâqin* –
haptomenos 227a29
mulâqin 545,19, *mutalâqin* 545,17 –
haptomenos 227a12
turning – *ḥaraka dawriyya* 753,23 I:
dawarân – *dinêsis* 243a25

uniform – *mutasâwin* 567,12 –
homalês 228b16

when – *matâ* 512,9 – *pote* 225b6
where – *ḥayṯu* 512,13, *ayn* 529,7 I:
ḥayṯu – *pou* 225b6; 226a25

Index of Names and Subjects

Achilles: Zeno's 'Achilles' problem 722,6
action: no motion in the category of 513,13
affection
  'sameness' or 'oneness' 561,2
  form is not an affection 498,10
Alexander of Aphrodisias
  infinite divisibility of changing things 649,24
  infinite divisibility of motion 658,22
  in succession 543,11
alteration 529,15; 530,6; 766,12
amber: attraction by 755,24
ancestors: can their number be infinite? 523,12
apart 538,18
attraction 755,24
  by a magnet or amber 756,5

beginning: of a change (motion) 675,13; 676,16; 677,16
between 537,9; 539,13; 545,2

capacity: no alteration in 766,15
carrying 753,19
categories: in which motion exists 490,9; 512,8; 529,7; 536,19
change
  accidental, partial and essential 491,2
  beginning of a change 675,13
  is between contradictories and contraries 501,1
  what changes is infinitely divisible 649,12; 650,9
  end of a change 667,8
  no part of something changes first 678,12
  instantaneous change 649,24
  see also motion
coming to be: see generation
comparable 783,9; 784,20; 787,7;799,8
condition 766,14; 767,6
  see also state
consecutive: see succession, in
contiguous 537,10; 544,12
continuous 537,10; 549,11
  magnitude, motion and time are continuous (infinitely divisible) 606,8; 613,4; 619,16; 636,15
  motion 540,9
contradictory 501,1
contrary 501,1
  contrariety in motions and states of rest 579,11; 586,3
  natural and non-natural motions as contraries 593,15
corruption: see perishing

decrease: see diminution
diminution 531,4
disposition 766,14; 767,7
  see also state
divisible
  what changes is infinitely divisible 649,12; 650,9
  motion is infinitely divisible 654,9
  magnitude, motion and time are infinitely divisible (continuous) 606,8

end: of a change (motion) 672,2
equivocal: see homonymous

form: no alteration in shape or form 766,12

generation
  comparability 799,8

is a change between contradictories 504,9
is not a motion 490,11; 503,11; 506,2; 506,20
what is generated has been generated before 692,10
no generation of generation 522,17; 525,2; 526,5; 527,9
growth 531,4

halting: not every state of rest arises by halting 599,7
having 512,9
homonymous 498,12; 784,21

immovable: see motionless
increase: see growth
indivisible: now is 640,13
infinite 628,5
an infinite number of ancestors possible? 523,12
possibility of infinite motion 730,n
instant: see now

knowledge: as an intellectual state 769,17

magnet 755,24; 756,5
motion 490,8
no motion in the category of action 513,13
is a kind of change 490,10; 510,12
comparability 783,9; 787,7
mover and moved must be in contact 753,16
contrariety 579,11
natural and non-natural motion as contraries 593,15
continuity 540,9
infinite divisibility 654,9
possibility of infinite motion 730,n
local motion 531,13
no motion of motion 515,11; 528,5
no motion in a now 646,6
no motion in the category of passion 513,13
no motion in the category of relation 513,7; 767,18
'sameness' ('oneness') of motions 551,14; 564,11
what is moved is moved by something 734,7
no motion in the category of substance 513,6
uniformity 567,12
no motion in the category of 'when' 513,3
see also change
motionless 534,13

now
is indivisible 640,13
no motion or rest in a now 646,6; 646,18

one: see same

passion: no motion in the category of 513,13
perishing: is a change between contradictories 504,12
Porphyry 530,7
position 512,9
pulling 753,26
pushing 753,24
pushing on 754,1
pushing off 754,2

quality 529,7
quantity 529,7

relation: no motion in the category of 513,7; 767,18
rest
as a contrary of motion 586,3
contrary states of rest 595,20
no rest in a now 646,18
rotation: may be infinite in time 730,n

same
'sameness' ('oneness') of motions 551,14; 564,11
'sameness' ('oneness') of states and affections 561,2; 561,11
shape
no alteration in shape or form 766,12
stadium
Zeno's problem of the 715,n; 722,13

state
  'sameness' or 'oneness' 561,11
  no alteration in 767,6; 768,14; 769,14
substance: no motion in the category of 513,6
succession: in 537,9; 543,6
synonymous 784,21

throwing 754,5
time
  is continuous (infinitely divisible) 619,16
  argument against ungenerated time 816,14
together 538,4; 538,6
touching 537,9; 545,17
turning 753,23

uniform motion 567,12

when 512,9
  no motion in the category of when 513,3
where 512,13; 529,7

Zeno
  his 'Achilles' problem 722,6
  his problem of the impossibility of motion 628,22
  his problem of the stadium 715,n; 722,13

SIMPLICIUS

On Aristotle on the Void

translated by
J.O. Urmson

introduction & notes by
Peter Lautner

Introduction

Peter Lautner

Simplicius comments on Aristotle's discussion of the void without writing a separate corollary on the subject, as he did for place and time. Consequently it is here that he has to face the objections to Aristotle by later philosophers, first of all by the Epicureans and the Stoics. In addition, it is impossible to disregard the impact that Philoponus' rejection of some aspects of Aristotle's conception had on his thoughts.[1] Simplicius' greatest effort was spent on responding to such criticisms. The quotations from the *Timaeus* at the end of our text, which serve to fulfil the obligatory task of aligning Aristotle with Plato,[2] are nothing but an appendix which lacks explanatory force. The word *phainetai* at 694,1 and 12 may bear witness to his own uncertainty about the claim in this particular case.

According to Simplicius, some of those who affirmed void considered it to be separate from bodies, but others held it to be interspersed in bodies. Accordingly, the proofs for its existence are divided into two groups. The arguments from motion and growth are put forward in favour of the separate void, the arguments from compression and nourishment are for a void interspersed in bodies.

In the following it may be useful to point out the ways in which Simplicius goes beyond the framework established by Aristotle. Although he agrees with Aristotle's conclusion that there is no void, he does not endorse all of Aristotle's arguments, and particularly not his arguments against the possibility of motion in a void. There are three crucial points in the treatise: (1) Simplicius' reply to the Stoics who had attacked Aristotle's reservations about extracosmic void, (2) His unreliable reply to Aristotle in defence of the idea of motion through void, and (3) his complaint that Aristotle does not sufficiently recog-

1. It remains unclear whether Simplicius had thorough information about Philoponus' commentary on the *Physics* (*CAG* XVI-XVII); what is sure is only that he was acquainted with the *De Aeternitate Mundi contra Aristotelem* (cf. *in Cael.* 135,27), which has survived in fragments. The alleged date of Philoponus' *Physics*-commentary (517, based on the evidence at 703,16-17 *CAG* XVII), or of its first edition (K. Verrycken, in R. Sorabji (ed.), *Aristotle Transformed*, London and Ithaca NY 1990, ch. 11), would make it possible.

2. For this requirement, see Simplicius *in Cat.* 7,23-32.

nise that the ground for the natural motion of bodies, whether in a void or not, is internal.

(1) Aristotle's objections to a separate void turn partly on the idea that, so far from facilitating motion, it would make motion impossible.[3] Void contains no differentiation to supply a sufficient reason for motion to occur in this direction rather than that, or (once it occurred) for it to stop here rather than there. Further, the yieldingness of void might make the cosmos disintegrate in every direction.[4]

In his discussion, Aristotle equated the centre of his spherical universe with the downward direction and the periphery with the upward. The Stoics agreed, but at the same time insisted on the existence of extracosmic void. They countered Aristotle's threat of cosmic disintegration by postulating a certain centripetal tendency in the universe, which prevents the whole substance (*ousia*) being scattered into the surrounding void. This inclination provides a holding power (*hexis*) that keeps things together and so retains their unity. This cohesive force seems to belong to two of the four elements, fire and air. In their cohesive role, they are called *pneuma* and they permeate the other two elements.[5] Through the fact that the centripetal inclination (*neuein*) is lodged in *pneuma*, it becomes suitable for saving the integrity of the cosmos as a whole. Periodic conflagrations (*ekpurosis*) come about when the other three elements are converted into fire. The theory has come down to us in its fullest form in the work of Cleomedes, philosopher and astronomer of the second century AD, but in some texts there are testimonies ascribing it to Chrysippus.[6]

Simplicius' knowledge of this doctrine is second-hand. He probably had no access to the original and it is not certain whether he knew of Cleomedes and his work at all. In his surviving commentaries there is no explicit reference to this author. He quotes only Alexander's paraphrase and takes over the criticism found there.[7] According to this, the holding power will do nothing to insure that the whole cosmos rests and is not carried away.[8] Alexander accepts that, by presupposing a cohesive force, the Stoics have managed to save the

3. See R. Sorabji, *Matter, Space and Motion*, London and Ithaca NY 1988, ch. 9.
4. The problem disquieted Strato of Lampsacus as well, cf. Fr. 50-2 (Wehrli).
5. See R. Sorabji, *Matter, Space and Motion*, ch. 6, pp. 85-9.
6. Cleomedes, *De Motu Circulari Corporum Caelestium* (H. Ziegler), Leipzig 1891, 10,15-12,5. The new edition by R. B. Todd is available in the Teubner series. For the view that Cleomedes' source was Chrysippus, see R. Goulet, *Cléomède: Théorie Élémentaire*, Paris 1980, 184-6, and R. B. Todd, 'Cleomedes and the Stoic concept of void', *Apeiron* 16 (1982), 129-36, esp. p. 33 n. 4. A detailed discussion of this theory has been offered by R. Sorabji, *Matter, Space and Motion*, ch. 9. For traces of this view in Chrysippus, cf. Plutarch, *de Stoic. Rep.* 1054E-1055A, Alexander, *Mixt.* 217,27-218,10.
7. 671,4-15.
8. 671,11-12.

unity of the cosmos (i.e. they successfully argued against the disintegration of the cosmos, 671,8-11). What he doubts is that the cosmos itself will be at rest, for this cohesive force is able only to hold the different parts of the cosmos together, but not to stop the cosmos allegedly moving to and fro (671,11-12). But it is a pity that Alexander does not mention Cleomedes' success in solving one of Aristotle's other objections.[9] Cleomedes makes room for motion in the void to have distinguishable directions *within* the cosmos by accepting Aristotle's equation of the cosmic centre with the downward direction.

(2) Another of Aristotle's objections to motion through the void is that a body moves faster or slower because the medium is different or because the body is more or less heavy. The medium can impede or help the body to traverse it since it becomes an obstacle when moving in the opposite direction or when at rest. And the denser medium impedes motion more effectively. There is a constant ratio which relates the time taken to traverse a given distance to the density of the medium. But void cannot stand in proportion to the full.[10]

These arguments prompted Simplicius to propose that all such absurdities are to be avoided by assuming that bodies move through void more quickly not only in virtue of traversing a medium more easily divisible, but also in virtue of their own power (*oikeia dunamis*).[11] This power is not explained directly in terms of the impetus, which in Philoponus' theory is implanted in projectiles, but faint traces may show that some such idea was in Simplicius' mind.[12] For elsewhere he hesitates about why in the case of projectiles motion is impressed by the thrower on the air, as Aristotle claimed, and not (as Philoponus claimed) directly on the thrown javelin.[13] At any rate, the difference in speed between bodies moving through a void interval and those moving through a filled medium does not issue exclusively from the different nature of the media, but also from the bodies themselves. In the case of falling bodies, some have greater downward impulse (*rhopê*), some less. Up to this point, he is in line with his rival, Philoponus. However, the latter goes on to say that what is obviated by the absence of resistance in a void is not the need for time in its entirety, but only the need for *extra* time for the body to complete its motion.[14] In Simplicius we meet with a different interpretation

9. A possible ground for Alexander's failure is that he was familiar with Chrysippus' conception only. That may then have been refined by Cleomedes, but, if so, this correction remained unknown both to him and to Plutarch whose criticism (*de Stoic. Rep.* 1055A ff.) is very like Alexander's.

10. 215b27-216a12, Simplicius, 675,27-678,7, esp. 677,22-678,1.

11. 678,1.

12. 677,24.

13. 1349,26-9 to 267a5-7.

14. *in Phys.* 681,14-30, 690,34-691,5, see R. Sorabji, op. cit., p. 147.

which remains problematic. For if one adopts the belief that a body can travel through the void more quickly because of its greater impulse, this implies that the velocity of bodies cutting across a vacuum need not be equal, which in turn suggests that their speed through the void is not, as Aristotle threatened, infinite. For talk of different, but infinite, speeds would have been rejected by the Greeks. If so, the ratio constructed by Aristotle between speed and resistance, which led to the charge of infinite speed, must also be dropped. Simplicius, however, seems to have recoiled from this step.

(3) The third problem occupies Simplicius throughout the commentary on the *Physics* and he tries to resolve it in different, and sometimes even contradictory, ways. One version to be found in our text is embedded in the criticism of the Epicurean explanation of why bodies are moving in the void.

As we know, Aristotle queried the possibility of motion in the void on the ground that there can be no preferred direction in it, whereas natural motion involves preferred directions, different directions for different things.[15] Moreover there is nothing to divide in the void. Thus things moving in the vacuum will have the same speed, even if there is an imparity in their impulse. It is therefore not surprising that the next philosophers cited here are the Epicureans; whether they reflected consciously on Aristotle's arguments or not, their account fits well into the line of Simplicius' thought as representing the position occupied by one of the protagonists in this 'debate'. As in the case of Cleomedes, their views had been sketched by Alexander whose text Simplicius simply incorporated into his commentary.[16]

According to Alexander, the Epicureans think that bodies move through the void because of their weight. Consequently, so Alexander infers, greater weight should result in faster movement. To avoid puzzles concerning infinite speed, the Epicureans claimed that this movement cannot be infinitely fast; rather it has to be considered 'swift as thought' (*hama noêmati*).[17] Whatever the dilemmas inherent in this notion may be, Alexander does not here reply by expounding them, but by bringing up the Aristotelian thesis which says that motion involves division of the medium by the moving body. However it is unwise to speak of void as admitting such divisions.

15. 215a12.

16. 679,12-22,32-7. The library used by Simplicius writing this commentary seems to have been scantily stocked with Hellenistic material, as is shown also by other passages of the *in Phys*. where the source is Alexander: e.g. 467,1-4, 530,9-16, 671,4-15, 1299,36-1300,10. Simplicius explicitly complains of the lack of Stoic texts at his *in Cat*. 334,1-3. This is relevant to M. Tardieu's thesis that he was writing not in Athens, but at Harran (*Les paysages reliques. Routes et haltes Syriennes d'Isidore à Simplicius*, Louvain-Paris 1990, pp. 128-35).

17. *Ep. Herod*. 61.

Simplicius is more sensitive in this matter.[18] His response is an interesting blend of Aristotelian and Epicurean ideas, not a hopeless jumble of them. Earlier he said that by beginning with the natural downward tendency or impulse (*rhopê*) of heavy bodies one could prove that their internal nature is the ground for their travel.[19] In expounding this thesis, he acknowledges that the efficient cause of the natural movement of the bodies must be internal. To appreciate this more, one has to be reminded of Aristotle's famous dictum according to which everything which is moved is moved by something else. Consequently, there must be efficient causes or movers external to the bodies in motion, even if these external movers are admitted by Aristotle to be accidental in the case of moving elements: mere releasers of the inner tendency, which still plays a role of its own.[20] Simplicius' solution is based on the idea that there is no other ground for the natural travel of bodies save their innate impulse and the difference in impulse results in different speeds, independently of the obstruction caused by the medium. In his view, the Epicureans were right in positing weight as the sole cause of downward motion. Their mistake was attempting to use this view for establishing motion in the void. On the other hand, he complains (rather unfairly) that Aristotle accepted only external agents as efficient causes of natural motion and failed to count on internal powers. Weight as natural impulse, he says, has to be considered the ground for such motion since how otherwise could one take account of, e.g., the difference in speed? Though ease and difficulty of the division of the medium contributes to velocity of motion, division is not the ground for the impulse. Quite the reverse holds, the innate impulse is the cause for division.[21]

This is Simplicius' view here. In honesty, we must admit that elsewhere he shies away from explaining motion in terms of inner agents. He takes sides firmly with Aristotle, against Philoponus.[22]

Synopsis

645,20, Preliminaries.

647,10, Criticism of Anaxagoras' refutation of the void and short doxography.

649,1, Sets out the supposed proofs of void.

18. His account is at 679,4-680,9.
19. 663,31-2.
20. *Phys.* 4.1, 208b11; 8.4, 255a1-6, a28-256a3; *Cael.* 4.3, 310b24-311a9. See R. Sorabji, *Matter, Space and Motion*, ch. 13, esp. pp. 220-1 on the accidental movers.
21. 680,8-9. For this reason I think the contrast concerning this issue between Philoponus and Simplicius proposed by M. Wolff, 'Philoponus and the rise of preclassical dynamics', R. Sorabji (ed.), *Philoponus and the Rejection of Aristotelian Science*, London 1987, 84-121, esp. pp. 96-7, a bit overstated.
22. 287,29-288,16. For Philoponus, see M. Wolff, op. cit., passim.

649,1, Argument from motion and growth.
650,15, Argument from compression.
650,30, Argument from growth and nourishment.
651,9, Argument from the case of the ash.
651,23, Pythagorean argument: the heavens inhale the void.
653,1, Examines the various concepts of void.
657,4, Against the unified concept of void, that it is place deprived of body.
658,21, Refutation of arguments.
658,21, Against the argument from motion.
659,29, Against the argument from compression.
660,11, Against the argument from growth and nourishment.
662,14, Against the argument based on the case of ash.
663,9, Proves that such void has no *hupostasis*.
663,18, Against separate void.
663,16, Against such void as ground for change.
664,1, Against such void as place for change.
664,1, Against void as boundless.
665,30, Against void as interval.
666,15, There would be no change if there were separate void.
666,15, There would be no natural change.
668,8, There would be no forced change (projectiles).
670,4, Neither would there be rest.
671,16, Fourfold argument against motion in the void.
672,3, Motion in the void would be timeless (instantaneous).
673,5, Void would stand in a ratio to the full.
675,5, This ratio is what the full has to the full, and objects would traverse the void and the full in the same time.
678,8, On unequal speed.
680,23, Reduces the problem to that of the interpretation of bodies.
682,25, Against void interspersed in bodies.
682,25, On rarefaction and condensation or compression.
682,25, Expounds the arguments introducing void from the rare and the dense.
684,10, Attacks them. Arguments from motion.
686,17, On the transformation of one element into another.
687,33, Eliminates the void when explaining compression.
690,18, Proves how the difference in speed follows from the weight.
691,24, On interspersed void as 'potential'.
693,3, Concluding remarks. Parallels with the *Timaeus*.

Translator's Note

The translation follows the text of *Simplicii in Aristotelis Physicorum Libros Quattuor Priores Commentaria*, edited by H. Diels in *Commentaria in Aristotelem Graeca*, vol. 9, Berlin 1882, except as indicated in the Notes and Textual Emendations. The lemmata containing the text of Aristotle under discussion are indicated in the text only by the opening and closing words; the translator has added the intervening text in square brackets for the convenience of the reader. He has provided his own translation to ensure uniformity of terminology and, occasionally, to comply with the text used by Simplicius where that differs from the received text.

The translation of the many Greek terms for change is a major problem. Here 'change' and 'motion' are always translations of *kinêsis* in its wider or narrower use and are indicated by a superscript ^{k}. *Phora* and *pheresthai* are translated as travel. *Metabolê* is translated as transformation. There is a price to pay, but it is worth paying. Tense and mood sequence in Simplicius do not always conform to that of classical Greek or modern English. The translation sometimes follows his use.

The translator wishes to thank Professor Sorabji and his colleagues at King's College, London, and also Dr Peter Lautner, for their aid in improving and correcting the translation.

Textual Emendations

649,8	*legontos*; Diels *legontes*.
653,5	*autous*; Diels *autois*.
658,18	*ton topon*; Diels *ho topos*.
664,20	*menei*; Diels *merei*.
665,28	*sôma ti*; Diels *sômati*.
673,29	Comma added after *pheretai*; comma deleted after *kenou*.
679,16	*autois*; Diels *autais*.
680,32	*eiper mê eiê*; Diels *eiper eiê*.
683,26	*sumbainei*; Diels *sumbainein*.
685,1	*hupokeimena*; perhaps *huperkeimena*.
687,4	Omitting *hoti* with F.
687,5	Adding *ek* before *tês* with a.

SIMPLICIUS

On Aristotle on the Void

Translation

Simplicius: On Aristotle on the Void

> **213a12-22** In the same way one must regard it [as a task of the natural scientist to investigate the void[1] also, as well as place, whether it exists or not, how it is and what it is. For they both cause the same belief and disbelief, on account of people's conceptions. Those who say there is a void treat it as a sort of place and vessel, which seems to be full when it contains the bulk it can accept, but void when deprived of it, so that the void, the filled and place are the same thing, though their essence is different. The enquiry must start by collecting what those say who affirm the void and what those say who deny it] and, third, the common opinions about these matters.

At the beginning of Book 3 he proposed to discuss change[k] as being 645,22
indispensable for the natural scientist, adding that it seems that change[k] is impossible without place, the void and time. So, he says, it is clear, both on this account and because these are common to all things and universal, that one must undertake an examination of 25
each of them. So he discussed change[k] and, next, the unlimited, since change[k] seems to be either limited or unlimited, and, being continuous, is divisible without limit, like everything continuous. After this he discussed place.

But, since certain people said that the void was place deprived of body, and that place was a void interval receptive of bodies,[2] he 30
responded to those who said that place was an interval receptive of bodies in his discussions of place. Now he will respond to the postulate of the void itself, whether it be place or anything else, indicating now also that the discussion of the void concerns the natural scientist, because those who affirm the void conceive of it as a sort of place and like a vessel.[3]

He says that there are the same problems about the void also as there are about place. First, 'whether it exists or not, how it is and what it is'. Aspasius well maintained that 'how it is' means the same as 'what it is'.[4] For while enquiring about the existence of place also, 646,5
and what it is, he used the words 'if it exists or not, and how it is and what it is'.[5] But there seems to be no room for the question how and what it is in the case of the void, since he proves that the void is altogether non-existent. Now, if he does not raise the question of how and what it is with reference to the existence of the void, but with

reference to those who posit it, how and what do they posit it to be? Or, since they say that the void is the cause of change[k], while he says that matter, which he admits to be called the void, is the cause of change[k], perhaps he admits the existence of the void in a way. Also, concluding the discussion of the void, he said 'So much for the distinction of the way in which the void exists from the way in which it does not',[6] since it does not exist in the way that those who posit it say, but nothing prevents the void existing as matter.

Also he says that the void admits belief and disbelief similar to that about place, since those who postulate the void said that it was place deprived of body, and they also made the fact of things changing[k] their place the chief ground for belief in the void, as for belief in place. So those who said that place is a void interval, saying that the void and place are the same thing, find the same grounds for belief and disbelief bearing upon their thesis. He made it clear that he himself thought that the grounds for belief and disbelief were similar for this reason, pointing out that 'they who say there is a void treat it as a sort of place and vessel'. He well maintains that the same thing is called full when it has within it the body of which it is receptive, but void when deprived of it. The same thing is place *qua* receptive; so what is identical in substrate is both void and full and place, though these differ in definition. But it is clear that it is not the full that is identical in substrate with the void and with place, but the interval. It is this that is at one time full, at another empty. Also place is the same thing *qua* receptive, but not *qua* full, nor *qua* void.

But, he says, he who enquires whether it exists or not and what it is must examine first the arguments of those who say it exists, then the arguments of those who deny it. From these it will become clear whether it exists or not, while we support those who speak more truly by examining the common conceptions or opinions about place and the void and the interval unqualified. For it is his custom to use common conceptions as a criterion and to compare and test opposing views with reference to them.[7]

It is clear that the account of place is naturally prior to that of the
647,1 void, since place is more general than the full and the void. For one place is empty, another full, and the general are prior to the particular. Moreover place seems to be distinguished positively by being receptive, whether or not it contains something, but the void by privation; but everywhere privation is posterior to possession. Further, place and the void are not the same thing, for neither is place void since one place is full, another void; but the full and the void cannot be the same thing, nor is the void place *qua* void, but *qua* being receptive and suitable[8] in nature to be filled by a body.

213a22-b4 Those who try to prove that the void does not exist [do not disprove what people intend to mean by it, but what they mistakenly say, as do Anaxagoras and those who argue as follows. They establish that the air is something, squeezing wineskins and showing how strong the air is, and catching it up in syphons. But people intend the void to be an interval in which there is no perceptible body; but, thinking that everything that exists is body, they say that where there is altogether nothing, that is void, and thus that what is full of air is void. So it is not required to show that air is something, but that there is no interval other than bodies, separate (*khôriston*) and actually existent (*energeiāi on*), which separates all body so that it is not continuous, as Democritus and Leucippus and many of the natural scientists say, or which is outside the whole of body which is continuous. So those men do not come to grips with the problem,] but rather those who affirm the void.

The project being to report what those who affirm the void say, and what those say who deny it say, he first undertakes to exhibit those who affirm it and support them with argument, and to display the argument as plausible, as was his custom. But, before producing such arguments, he shows the errors of those who try to refute them but who oppose them badly. In this way he first introduces those who deny the void, while at the same time showing against what conception of the void the battle should be. It is not that which the generality use, who think that the only bodies are those that resist, and who say that the whole place of the so-called air is void, as if air were nothing.

Thus the followers of Anaxagoras think that they refute the existence of a void just by proving that air is something. So they would show that it was a body, and a resistant body, by compressing it in wineskins.[9] It is more appropriate to hear Aristotle saying how they brought out blown-up skins and displayed how hard it is to squeeze them when there is air inside, though it is very easy to squeeze them when not blown-up. They even exhibited water-takers, i.e. syphons, which do not accept water when containing air, but, when the air is sucked out, readily take up water and do not let it out until one takes away the finger stopping the hole and permits an amount of air to enter corresponding to that of the water going out. So those who attacked the void argued in this way against those who said that the place of air was void, because they were mistaken. But they did not prove that there was no void, for people think that the interval in which there is no perceptible body is void. And, since they think that 648,1
everything that exists is body, and suppose that what is not body is nothing at all, they say that the interval is void in which there is

nothing, since they have this conception of the void. But, as for air not being a body and therefore there being nothing in the interval containing air, this they say mistakenly. So it is not this that they ought to have opposed, since it is a mistaken view, but the very conception of the void itself, showing that there is no interval other than that occupied by bodies. For this is what 'nor separate' implies; for even if it be filled, but something other than body, it is separate from bodies. They should also show that no such interval exists, but is a mere conception,[10] which is what 'nor actually existent' implies. For these said that there was actually such an interval, present between bodies, and not permitting them to be continuous. This was the view of the circle of Democritus and Leucippus, who said that there was void, not only within the universe, but also outside it, which would clearly not be a place, but would exist independently.[11] Metrodorus of Chios shared this view,[12] and certain of the Pythagoreans,[13] as Aristotle will soon say.[14] Later Epicurus shared it.[15]

Porphyry, however, does not write 'neither separate nor actually existent', but 'neither inseparable from them nor separate'. 'For', he says, 'Democritus' circle made it inseparable, so that the whole is also not continuous, bodies being separated by the void. But those made it separate who said that the void was outside the universe, leaving the whole continuous, as did the Pythagoreans.'[16] But I think that the former reading is better, even though this is reasonable; for of those who expressed a view about the void some said that it was separate in its nature, pervading the whole of the cosmos and extending beyond the cosmos, continuous in itself. But others said that it was interspersed everywhere within bodies in small portions, in that way seeming to be inseparable. That is why, of the four arguments following, one of them introduces the first sort of void on the ground of motion[k] and growth, but the one from compression and ashes the second. Thus those who wish to attack the void on the ground that the air is something do not use the proper doors and entrances as they approach this notion of the void, but miss the doors, i.e. approach externally and through other doors, and as if trying to enter through a wall that gives no admittance. For the adage also is applicable to them. So these do not use the doors, but the others are more persuasive who say that the void exists, whose arguments he proceeds to set out.

649,1 **213b3-15** One thing they say is that change[k] of place would not exist [(this is travel and growth). For it seems that there would be no motion[k] if there were no void, for the full is incapable of receiving anything. For if it does receive, there will be two things in the same place, and it would be possible for any number of things to coincide. For it is not possible to say what difference would stop this happening. If this is possible, then even the smallest thing will receive the biggest; for a big thing is many little things. So if many equal-sized things can be in the same place, so can also many of different sizes. So Melissus also proves that the whole is immovable[k] on these grounds; for, if it will move[k], he says that there will have to be a void, whereas there is no void.] So that is one way in which they prove that there is a void.

He sets out four arguments of those who affirm the void, one using the concept of the separate void, and three using that of it as interspersed in bodies. He adds a fifth to them which is a doctrine of the Pythagoreans.

The first of the four is something as follows, starting from a disjunctive argument that says[17] that the change[k] of place that is seen both in local motion and in growth occurs through a medium that is either void or filled. But it is not through one that is filled, as will be proved; therefore it is through a void, and the void exists. That it is necessary for the motion[k] to be through some medium is seen as self-evident to those who realise that exchange of place from somewhere to somewhere occurs through an interval intervening between the places from which and to which, that are isolated from each other. It is also manifest that every interval is either filled or not filled; for the disjunction is of contradictories,[18] and what is not filled is altogether void. He also proves by a hypothetical syllogism that the motion[k] cannot be through the full as follows: if change[k] of place of bodies occurs through the full a body will pervade a body and 'the smallest will receive the biggest'. But surely it is impossible for the sea to be contained in a cup of water; so the antecedent that the motion[k] occurs through the full is impossible.[19] He proves the consequent that the smallest will receive the biggest from the fact that if the former body will receive any other body, so that it is possible for two bodies to be in the same interval, these will also receive a third which occupies the double body, and this whole will receive a fourth, and so on indefinitely. For nobody could say why a body had received one other but not more – for the one was as full as the two. Therefore it will receive also the biggest, since what is big can be divided into many [little bits] equal to the little substrate at the beginning. For, as he says, a big thing is many little things, since each of the little bits in the big thing that are equal in substrate will come into the original little thing, so that the whole would be in it and the smallest will receive the biggest, which is obviously absurd. It is clear that, if it were to receive what is equal to it, it would receive also the unequal; for more equals make an unequal. So it will receive unequals also, and so also the biggest. But even receiving the unequal is itself absurd, which is why he did not omit it.

He confirms the antiquity of this argument that introduces the void on the basis of motion[k] from Melissus' treatment of the consequent as an obvious one, saying that, if what is moves[k], it moves[k] through a void, and adding 'but, however, there is no void', and concluding that 'therefore what is does not move'. It is clear that Melissus thus in a way relies on the argument.[20] But he relies on it not with regard to the bodily or the partial, but to the intelligible and perfect. For he claims that this is one and unchanging[k], proving, I think, its unchangingness[k] through it being all and there being nothing beyond it, whither it will be transferred through the void. For THERE[21] there is no void, nor, perhaps, even diversity, since it is everything. Also what is not will have no place in what is wholly real (*to pantelôs on*). And, even if there is diversity (*heterotês*) THERE, by which the forms are distinguished from each other, still diversity is a reality. The void has no place in that which wholly is, as does that which is not. 'But do you guard your thought from this way of enquiry' as the great Parmenides says.[22]

213b15-18 They argue in another way that things visibly contract and are compressed. [They say, for example, that jars receive the wine with the skins, which implies that] the compressed body contracts into the voids within.

He sets out a second argument, the one from the compression of bodies. For if, the same substance remaining, bodies contract and expand, they say that it is clear that their collapse into themselves comes about because there are internal void intervals. The fact that the jar accepts the wine together with the skins is put forward as a confirmation of this. Alexander took this to mean that the jar, even if its complement of wine be put into skins, will receive it with the skins.[23] But perhaps this is excessive and caused by the unhappy composition, and it is that the same jar, being full of wine, will receive the skins capable of containing it if put in as well without overflowing. This, they say, is impossible unless the wine is compressed; but this would not have happened unless there were intervening void intervals, by the collapse of which the compression is achieved. Those who explain compression by the void also say that heavy and light things differ by the latter containing more void, the former less.

213b18-20 Also growth seems to all to happen because of the void, [on the ground that their food is body,] but two bodies cannot be in the same place at the same time.

Having said that change[k] of place would not occur if there were no void, and that of change[k] of place one sort is by growth the other by travel, and having made this point with regard to travel, he now says 651,1
how they used to prove that change[k] through growth would not occur if there were no void. For if growth occurs, food being distributed throughout, and food is body, it must occur either through the full or through the void. But it is impossible for it to be through the full, as has been proved, since body cannot permeate body so that two bodies coincide. So it is clear that the entrance of the food comes about into the voids of that which is nourished which are scattered throughout, and thus makes what is nourished grow on all sides. The words 'seems to all to happen' shows the argument to be based on common opinion.

> **213b21-2** They also call to witness the fact about ashes, if they receive as much water as does the vessel when it is empty.

This stands as the fourth argument in support of the void. It is in a way the same as that about the wine and the skins, since it is also based on the compression of bodies. For as the jar receives the skins in addition to the wine by the compression of the wine, so the pail full of ash receives the water in addition to the ash as the empty spaces in the ash contract. But this illustration differs from the one with the wine, I believe, since in the former case the same receptacle received in addition the skins which did not have the same bulk as the wine, but here both the ash and the water are sufficient on their own to fill the vessel, clearly through the collapse of the pores in each, or, rather, those in the water collapsing, but of those in the ash some contracting, some taking in the water.

> **213b22-7** The Pythagoreans also said that there was a void, [so that it enters the heavens from the infinite breath, since they inhale also the void which distinguishes different natures, the void being an agent of a certain separation of a series, and distinguishing its members. This is primarily in the case of numbers,] for the void distinguishes their nature.

Using the doctrine (*endoxon*) of the Pythagoreans, he adds their opinion as a fifth argument for the existence of the void. For these said that the void entered the cosmos which, as it were, breathed in or inhaled it like a breath from that which surrounded it outside. It fulfilled a need to prevent all bodies from being continuous with each other, as Alexander understands them.[24] But Aristotle did not understand them as referring to bodies, but, he says, it 'distinguishes different natures, the void being an agent of a certain separation of a series and distinguishing its members'. For the members of a series,

652,1 with nothing between them, are what the void distinguishes; things separated by other things between them are not separated by the void but by those things. Such a power of the void applied to numbers and appeared first to distinguish their natures. For what else is it that distinguishes the monad from the dyad and this from the triad except the void, since no substance was between them?[25]

But what might be these riddles of the Pythagoreans? Is it that the otherness that distinguishes the forms THERE beyond the bodily cosmos was participated in by the cosmos and so brought about the distinction and separation of the forms in it, there being no void THERE (for the beautiful, for example, is different from the just, not because it is not just, but because everything is in accordance with the beautiful through the union THERE, and because in that which wholly is there is not that which is not).[26] But HERE a separation comes about through the intervention of that which is not. For the monad is not a dyad and the dyad not a monad, and the non-existent between them is the void which separates the forms in the cosmos, just as the otherness THERE beyond the cosmos does the forms. It is a being itself also and is not called not-being, and, therefore, not void, but is the cause of the void HERE. That is why Plato in the *Sophist*[27] called it too, in a way, not-being.[28]

These, then, are the arguments of those who said that there was void as set out by Aristotle. But Strato of Lampsacus reduced the four to two, that from change[k] of place and that from the compression of bodies, but adds a third which is from attraction.[29] For it happens that the iron-stone attracts other iron things through yet others, when the stone draws out the contents of the pores of the iron, with which material the iron is also pulled along, and this in turn draws out material from another, and this from another, and thus a chain of pieces of iron hangs from the stone.

213b27-9 The arguments of those who affirm and those who deny the void are roughly these, in character and quantity.

He set out several arguments of those who affirmed the existence of the void, and, before these, the argument of those who abolished it, the followers of Anaxagoras. He said that this was beside the point since it proved that air was a body, not that there is no void.[30] He thus now reasonably says that both lots of arguments have been set out, those affirming the void and those denying it.

653,1 **213b30-214a6** To determine which view is right we must settle what the word means. [The void is thought to be a place in which there is nothing. The reason for this is that people think that what is is body, and every body is in a place, and the void is the place in which there is no body, so that, if somewhere there is no body, there is nothing there. They further think that all body is tangible; of that kind is that which has weight or lightness.[31] So it follows syllogistically that that is empty where there is nothing heavy or light.

This, as we said previously, follows syllogistically; but it is absurd if a point is void,] for void has to be a place in which there is an interval within tangible body.

He has set out together those who deny and those who affirm the existence of the void. He reasonably says that the time has now come to examine them,[32] to see which answer to the problem[33] is right. But since those who argued against it, and perhaps those who argued for its existence, appeared to have a vague conception about it, he therefore says that first one must determine 'what the word means'. It is clear that this problem is also prior to that of what it is, as we were taught in the *Posterior Analytics*.[34] For we cannot investigate whether something exists of which we have not already formed a concept, just as we cannot investigate what it is unless already convinced that it exists. The meaning of its name and what it is are different, since the former is gathered from the character of our conceptions about it, whether what it signifies exists or not, as will now be seen in the case of the void, although it does not exist, but what it is is gathered from the character of the thing itself. Therefore that of which we investigate what it is must definitely exist. Also, the problem whether it exists is prior to that of what it is. Alexander says: 'As in the case of place he first expounded the meaning of "being in something",[35] so here he asks the meaning of "void".' But perhaps in the case of place, because of agreement about the concept, he did not need to start with the meaning of the word, and it was not for this that he started with an enumeration of the meanings of 'being in something', but it was in order to distinguish both the genus of what was in place and being in place from the other sorts of being in something, and especially being in something as a whole. So he says first what the word 'void' seems to mean, that it seems to mean in common use 'a place in which there is nothing'. He adds the reason for this opinion, and not that it is where there is no body. For since the naturalists, or those who have never contemplated bodiless being, say that everything that is is body, they clearly say that what is not body is nothing.[36] So it is the same to say 'in which there is no body' and 'in which there is nothing'. Nor does this suffice for them. But, as they say that everything that is is body, so they also say that all body is tangible.

That is why they said that air was neither a body nor an existent. So, since everything is tangible that has weight or lightness, by a syllogism they come to say that the void is 'that in which there is nothing heavy or light' – consequently that in which there is air.[37] Also, since they would speak so vaguely in positing in what the void consists, it follows that they call a point and a line and numbers void.[38] For they contain nothing heavy or light, whatever else is contained in them. But in this way they depart from the concept of the void. For it is supposed to be a place and an interval, 'where there is an interval between tangible bodies'. But it is not then void when there is body in it, but when it is such that body can be in it. But a point is not like that. It is clear that the absurdity was brought about from the vagueness of the statement wherein void consists, since they did not also make it clear that it must be some interval.

So Aristotle despises the complacency of those who form such hypotheses, and he turns to the concept of the void, and from now on hypothesises it to be an interval.

> **214a6-11** So, then, the void seems to be described [in one way as that which is not filled with body perceptible by touch. But that is perceptible by touch which has weight or lightness. So one might raise the problem what they would say if the interval contained colour, or sound – that it is void or not? Or is it clear that, if it would receive a tangible body, it is void,] if not, not.

Having spoken of the absurdity that follows from the vaguely stated meaning of 'void', he adduces those who describe the void as 'that which is filled (or not filled) with body perceptible to touch'. Both readings are found.[39] This is taken as an interval as is also shown by the aporia that follows: 'if the interval contained … .' If the correct reading is 'not filled with body perceptible by touch', it would mean an interval capable by nature of receiving a body perceptible by touch and being filled by it, but which has not received it or been filled. But if it is 'filled with body perceptible by touch', in this case also one must understand it as 'suitable to be filled', not as 'actually filled'. He made this clear by saying, 'Or is it clear that, if it would receive a tangible body, it is void?' Whichever the reading, the absurdity adduced is as follows: if that interval did not contain a body perceptible by touch, but was supposed to contain a colour, or a sound or some other quality perceptible by some other sense, is that a void or not? For if it were called a void, how could that be called void that contained a perceptible quality? But how could a void be perceptible? But, if it is not void, it will be something not void but not filled by body perceptible to touch. Observe that that interval containing a colour or sound and suited to be filled by a tangible body, though an interval and not filled, will still be said not to be a void. But he himself who stated the absurdity added the solution that they required as coming from them, saying, 'Or is it clear that if it would receive a tangible body it is void, and if not, not.' This is if we do not allege that this absurdity is that this interval would be called void, even if capable of receiving a tangible body, but did not contain it, but a colour or sound or generally a perceptible quality.

But Alexander says: 'If the reading is "that which is filled with body perceptible to touch", it can mean that which is perceptible by touch alone, and not by any other sense. It is like that to say that air is a body perceptible to touch, since there was a belief about the void saying that air was void, though tangible through its having an impulse of weight or lightness. The circle of Anaxagoras tried to overturn this belief by squeezing wineskins. If', he says, 'it was said about air, the problem would be as follows: if this air, that they call void, were to gain a colour or a sound or some quality perceptible to some other sense in addition to touch, would it still remain void or not? If they were to say that it was still void, it would not be that filled with the perceptible to touch that was void, as was said, if it still remained void when it had other perceptible qualities. But if the air was no longer void, having added one of these, the void would be receptive of qualities that were not of bodies. In that way there would be in a place qualities, but not bodies.'[40] But Alexander himself seems not to accept this interpretation. However, he says that what comes next, 'Or is it clear that if it would receive a tangible body it is void, and if not, not' does not apply to air and its sort of void, but to the interval. For those who said that air was void also said that the interval in which there was air was void, on the ground that air was nothing, so that, if they had thought that air was a body perceptible by touch, they would not have said that the interval containing it was void. Nor would the followers of Anaxagoras have been attacking the void by showing that air was a body and powerful. That was why the void was defined as an interval suited to be filled by a body perceptible to touch, so that what was receptive of air could be seen as void, air not being a body perceptible to touch, as they thought.

Alexander says that there is alleged to be yet a third reading which says 'that which is filled by a body imperceptible to touch'.[41] The interval would be an imperceptible body, if it means that which has three dimensions and is therefore called body and full, since the definition of a body seems to be 'that which has three dimensions'.[42] But Alexander says that the solution given does not fit this reading either; I do not know why, since the void is taken as an interval. But the proposal to say that the void was an interval filled with body imperceptible to touch was consistent. But Alexander says that by its nature air does not share perceptible qualities perceptible to other senses, but is only tangible. For it would not serve the other perceptions which occur through it, if it also exhibited some quality belonging to them. For, if it were, perhaps, black, how would it serve for the transmission of white?

It is clear that according to that account of the void which says that that is void in which there is no body perceptible to touch, i.e. heavy or light, then, if there be some interval in which are the heavens, that would be void. For the divine body in circular motion is neither heavy nor light, as was proved in *On the Heavens*.[43]

But it seems that the first absurdity adduced – that a point is void 656,1 – comes about because of the vague meaning of 'in which', and this because of that which does not exist in a void being distinguished by touch only, and being said to be void for that reason.

214a11-16 In another way [that is said to be void in which there is no particular thing nor bodily substance. That is why some say that the void is the matter of body – those who also say the same about place – mistakenly. For matter is not separable from things,] but they are looking for a separate void.

He hands down another conception of the void, which says that that is void in which there is no bodily substance.[44] For in explaining 'particular thing' he added 'nor bodily substance'. This would differ from the previous meaning since there that is said to be void where there is no body perceptible to touch, which is characterised according to the resistance of that body which they failed to see in the air and so said that an interval was empty in which there was air. But here that is said to be void in which there is no bodily substance, while a bodily substance has not only weight and lightness, but also colours and shapes and magnitudes. So according to the previous account the interval in which the heavens are would be void, because the heavenly body has neither weight nor lightness;[45] but according to this account it would not be void, since that is void in which there is no bodily substance, but the heavenly is a bodily substance. Also, perhaps in fact the former account was by those who did not regard air as a body, because it was not resistant, and therefore thought that the interval was void in which there was air because it was not resistant. Attacking these, Anaxagoras was able to show that air was a body. For if they said that that was void in which there was no resistant body, but air, even if not resistant, is still none the less body, and where it is is not void, it is clear that what they spoke of would not now be said to be void.

But having posited those who say that the void is that where 'there is no bodily substance', and in that way seemed to speak more accurately than those who add 'tangible' to 'body', thus falling into absurdity, he drives these also into saying that the void is matter which is simply such, since it is this in which the form of body comes to be. So those who have this conception of the void say that matter is the void.[46]

These same people also say that matter is place, just as they say that the void also is place, so that they say that the void, matter and place are one in substrate. Earlier on, also, in the discussion of place, he refuted those who say that matter is place.[47] But now, adducing those who say that matter is the void, he confronts them also with the same absurdity in saying that matter and void are the same, comparing the opposite conceptions of them, in which matter is not separate from things of which it is the matter. But those who affirm the void seek for it as separate from that body which is of a nature to 657,1
occupy it, so as to prove here also in the second figure[48] that the void is not matter. But he admits that place is in a way designated by the word 'void'.

> **214a16-26** Since place has been defined, [and the void, if it exists, must be place deprived of body, and since it has been stated both in what way place exists and in what way it does not, it is clear that in this way there is no void, whether separable or inseparable. For the void is supposed to be not body but the interval of a body. So this is why there seems to be a void, because there is also place, and for the same reasons. For change[k] of place comes to the aid both of those who say that there is place apart from the bodies that enter it and of those who say that there is void. They think that the void is a condition of motion[k] as being that in which there is motion[k];] this is like what some say about place.

Having set out what seem to be the different conceptions of the void, he has suggested that all tend towards one and the same thing – that the void is a place deprived of body. For whether it be a place where there is nothing, or an interval not filled by body perceptible to touch, or that in which there is no particular thing nor bodily substance, by all these accounts the void is a certain interval and a place deprived of body. For we think of place as a place then when it has received a body, since it is the place of something, but as void when it has not yet received a body. But to be intervals separate from bodies holds of both of them, place proper and the void.

So, since it had been shown of place that it is the limit of the container and is not an interval as such, it was then proved by an argument parallel to that about place that the void is not an interval. But now, since we are faced with an examination of this very matter that remains after the meaning of the word 'void', whether the void, i.e. the above-mentioned interval, exists at all, it remains to show that there is no such thing, neither as not having body within it, which is what 'separated' means, nor as having it, which is what 'unseparated' means. For of those who affirmed the void, some spoke of it as being something on its own, either never containing a body, as the void outside the universe, or sometimes doing so, as that in rare things and that which separates bodies so as not to be continuous. But some spoke of this interval as having some nature of its own, but as always filled with body. But 'separated' can mean that said by them to be outside the universe and 'unseparated' that carried about in bodies, and not that which always contains body. For this hypothesis was clearly common in Platonic circles after Aristotle's time.[49]

He shows that the arguments stated against place being such an interval could also be satisfactorily stated against such a void, since the same ones were in favour of such a void as of place – that it was receptive and separate, and that there could be no change[k] of place.[50] For they clearly intend the void to be not a body but an interval receptive of body. So there seems to be a void for the same set of reasons as there seems to be place.

But Alexander says: 'By saying that the void, though three-dimensional, is not body, one can refute those who wrongly think that in 658,1 the *Categories*[51] he means by a quantity of body a mathematical interval on its own without matter. For he nowhere calls such a thing simply body, but either the interval of a body or a mathematical body; but there he simply called a body a quantity. It is clear that the mathematical one would not be the interval of a body. For here also the void is said to be the interval of a body, as being an interval receptive of a body, but not a bodily interval.' Alexander, so interpreting, argues that if it is not a body, but the interval of a body, and if it is impossible for there to be such an interval without a body, it is also impossible for there to be a void. But perhaps 'the interval of a body' means the three-dimensional, such as is a bodily interval. Then he would be saying: if everywhere the three dimensions are of a body, those who posit the mere interval of a body are positing that which is not.[52]

Since, as has been said, there are many grounds for the inclusion of place among realities, and also of the void, he pointed out their multiplicity by saying 'and for the same reasons'; but he set out the most important. 'For', he says, 'change[k] of place comes to the aid both of those who say that there is place apart from the bodies that enter it and of those who say that there is void.' But, since it is possible for there to be many different grounds for the same thing, he showed the void and place[53] as being the same ground of change[k] of place, saying that those who posited the void called it a ground 'as being that in which there is motion[k]'. Place was a ground in the same way.

214a26-32 But there is no necessity [for there to be a void if there is motion[k]. By no means in the case of all change[k] in general, for a reason that escaped Melissus also. For it is possible for that which is filled to suffer alteration. But neither is it so for change[k] of place. For things can simultaneously make way for each other, there being no separate interval beyond the moving[k] bodies. This is clear also in the rotations of continuous objects,] as it is in those of liquids.

Having determined the meaning of the word 'void', that it is an interval and place deprived of body, his remaining project is to show that this does not exist, whether separate or inseparable. First he shows that the arguments in favour of the void have no force. First, that which was first put forward, which seems to be more weighty than the others, which said that if there is motion[k] there is void. Now he shows that it was argued without force, a little later that quite the reverse holds – if there is motion[k] there is no void, and if there is void there is no motion[k]. Now, in the meantime, he attacks the argument through its open-endedness; for even if the void is a ground of change[k], it is not so of all – for it is not so of alteration. Something may grow hot and turn white and generally be altered when it is filled, and has no need of the void. Also he criticises Melissus for saying open-endedly that the whole is unchanging[k] because there is no void.[54] He seems to have been unaware that nothing prevents there being alteration, even if there be no void. It has already been said what kind of whole Melissus said was unchanging[k] because of there being no void.[55] He ought, then, to have claimed, if anything, that 'if there is change[k] of place there is a void', and not simply 'if there is change[k]'. He censured this, which one might call an oversight by Melissus in Book 1 also,[56] saying 'further, why should there not be alteration?'

So, having dissolved the consequence through the open-endedness of the postulate, he goes on to show that the void is not even a ground of change[k] of place. For bodies may exchange places with each other and thus change[k] their place by taking over each others' places even without any void of the sort he called a separated interval, i.e. one not in a body but on its own. He showed the possibility of this by the case of continuous bodies revolving and always staying in the same place. Tops, for example, change[k] their place without needing a void. Water, also, when it moves[k] by swirling round in a vessel, maintains the same place as a whole, but its parts exchange places with each other, changing[k] their place. Neither does it take over a void, nor does body occupy body. He also used this example in Book 1 when confronting the changelessness[k] of Melissus.[57] But perhaps one might say that such things as retain the same place relative to the universe do not change[k] their place relative to us, given that we say that the whole heaven does not change[k] its place. Thus the void will seem not to be a ground of change[k] of place only for those things that as wholes remain always in the same place. So Strato's illustration is more suitable, since it avoids these doubts. For if one throws a pebble into a vessel full of water and turns it upside down, stopping outflow at its mouth, the pebble will move to the mouth of the vessel, the water taking over the place of the pebble. The same happens in the case of swimmers, whether fishes or anything else.[58] Aristotle himself will have more to say about this exchange of place.

214a32-b1 Condensation is possible, not into the void, [but by extruding contents], e.g. the contained air when water is squeezed.

The second argument of those purporting to show that the void exists was that from compression. The example of this was that from the skins, which the jar receives together with the wine through the compression of the wine, which is merely its congregation into the contained voids. In reply to this argument he says that it is not necessary that what is compressed should be densified by having voids within itself into which it is gathered. For it is possible that the remaining parts should congregate into a lesser volume by some lighter body being extruded and squeezed out. The compression comes about by the departure of the lighter and thinner, such as air, fire and water, and, generally, of the rarer from the denser, as the denser congregates. Accordingly, the compressed body is not only smaller in volume but denser in its whole substance. The verb 'squeeze out' [*ekpurênizein*] derives from those who shoot off olive-stones [*purên*] by compressing them with their fingers, because of their tapering shape.

214b1-3 Also growth is possible not only by something coming in but also by alteration, e.g. if water comes from air.[59]

The third argument was from growth, proving that the void exists by such a hypothetical argument as 'if there is growth there is void: but the first, so the second'.[60] It confirms the hypothetical by the fact that the food is added to the feeders by travelling through a void, since body does not pass through body. He says that it is altogether unnecessary, if something grows by the entry of some body, that it should grow by accretion. For many things grow and become greater in bulk after being smaller, not by the accretion of something, but by alteration and changing into one of the bodies with lighter elements, as when air comes from water. For the bulk of the air that has come from the water is greater than that of the water from which the air came. So the proposition 'if there is growth, there is void' is not true.

In this context there is nothing remarkable in calling the change from water into air alteration, though it is the coming to be of air and the passing away of water. But the matter of growth is worthy of further consideration. First, then, how does he say growth occurs in the case of the inanimate, and how does he say they grow without food, though he himself says clearly in *On the Soul*[61] that only things that feed grow, and it is the animate that feed? For he says that only those things that have a nutritive soul feed and grow. How does he now say that things grow without the entry of food, in order to escape the absurdity urged on both sides of the question whether food travels through the full or the void?

In answering these objections Alexander says: 'Since it is not yet 661,1
clear how accretion comes about in things that feed and grow, and what it is that grows, whether form or matter, about which he will enquire and make distinctions in *On Generation*,[62] and will show how growth comes about, and what it is that grows, whether form or matter, for this reason he now gives growth a more general sense, and thus attacked the argument from it to the void.' But if he was dealing with growth in general and food, how was it reasonable to do so missing the point, even if he has not yet given instruction on these questions? Perhaps, then, as he first replied in the argument about change[k], that they were saying that the void was indefinitely a ground both of all change[k] and of change[k] of place, so now also he first replies as taking them to mean by growth any kind of increase, and shows that it is possible to acquire a greater size even without transmission of food, and then, next, in the same way, as we shall learn, he supposes them to mean growth proper through addition of food, and to raise the puzzle on this basis, and he tries to turn the puzzle round against them, saying:[63]

214b3-11 But, in general, the argument about growth [and the water poured into the ash is self-defeating. For either not any and every part grows, or not by body, or two bodies can coincide (so they require the solution of a problem common to all, but do not show that the void exists), or it is necessary for the whole body to be void, if it grows everywhere and not through the void. The same argument applies to the case of the ashes.] So it is clear that it is easy to refute these grounds for showing that the void exists.[64]

If one understands 'growth' indeterminately as applying to things increasing to a greater size, it is sufficient to confront them with the change of water into air and, in general, from the less in volume of body to the greater. But if we understand it as applying to growth proper, which comes about by the accretion of food, those who introduce the void because of the entry of food will no longer be overthrown by other arguments, but, as in the proverb, will be caught by their own wings,[65] or, as Aristotle more literally said, 'the argument is self-defeating'. For, in order to solve a problem common to all, that of food, it introduces the void, which does not solve but rather exacerbates the problem. For if the accretion occurs through the food travelling through the void, and the growth is thus achieved, as they say, either the food does not come through body (and what would be the need for the hypothesis of the void, which they posited because body does not travel through body, food through the fed?), or, if food is body, either it will not be fed and grow in every portion (which is contrary to fact, for bodies are fed and grow in every part), or if it be fed and grow throughout, then either body will travel through body – to avoid this they posited the void – or the whole body will be void, if the whole of it grows, and growth will be through the void, so that the body will no longer have void in itself, but will be itself void; so body and void will be identical.[66]

So those who posited the void do not prove that there is void, but in trying to solve the problem of food that is common to all by means of the void, they make it more intractable. By inserting within the list 'so they require a solution of a problem that is common to all but do not show that the void exists' and then adding the alternative 'or it is necessary for the whole body to be void' he introduced a little unclarity into the passage.

He solved these puzzles also in *On Generation and Corruption*,[67] where he raises the problem of food, saying that there is not accretion in every part whatsoever, but something runs out and something is added. When what runs in is more than what runs out there is growth. Similarly when what runs out is more than what runs in there is decay. For the food comes into the channels of the body and grows into it by assimilation. But the whole of body is not channel, but, where there is now channel, that becomes full, and what is now full becomes a channel by outflow.

But since there was also a fourth argument which introduced the void on account of ashes, because the vessel full of ash receives the same amount of water poured in with the ash as it would receive when empty (for, they say, either body will pass through body, which is absurd, or the water will pass into the voids in the ash), he says that the problem rebounds on these people also. For in avoiding the passage of body through body they will be compelled to say that the ash is incorporeal and void throughout, in order that the vessel should receive the same quantity of water as if it contained no ash. For if, being body, the ash should receive the same amount of water, it will thus follow that body passes through body, to avoid which they posited the void. So these people also, in trying to solve the problem by the void, will be hindered by it from solving it.

Eudemus solves the problem of the ashes in Book 3 of his *Physics*, saying that 'what they refer to can happen without voids. For some heat seems to be included in the ash as in lime. It is clear: when water is poured in, both of these burn, the lime itself, and, in the case of the ash, the water filtering through the ash burns the bodies. As this goes on, much vapour is given off, so that the volumes are decreased because of the vapour.'[68] But, even if it is diminished, the whole of the substance of the ash is surely not spent. So I think that it is possible to come to the aid both of those who posit the void and of those who posit compression, saying that it is not the voids in the ash alone that solve the problem, but that also the voids in the water close up, pressed by the filling of ash, so that it does not have the same 663,1
volume as the water on its own that is poured into the ash.

These then are the replies of Aristotle to those arguments he reports as positing the void. But Strato, giving his answer to the problem of attraction also, says: 'Nor does attraction require the positing of the void. It is not even clear whether attraction exists, for Plato himself seems to refute the attractive power, nor, if attraction does exist, is it clear that the stone attracts through the void and not through some other cause. Those who say it does posit a void, they do not demonstrate it.'[69]

> **214b12-17** Let us repeat that there is not such a separated void, [as some say. For even if there is some natural travel of the simple bodies, as for fire upwards and for earth downwards and towards the centre, it is clear that the void would not be the cause of the travel. Of what, then, will the void be the cause?] For it seems to be a cause of change[k] of place, but of this it is not.

Having shown that of the arguments that seem to establish the void some have no validity, and those that seem to solve the problem fall into a greater, he set out finally and pre-eminently that there is no foundation for such a void. So, since, as was said earlier, some say that there is some separate interval on its own, a boundless void that is a ground of the change[k] of place by bodies, some posit a void that is interspersed and mixed with bodies, he meets the former opinion first. He says that if their reason for saying that there is a separate void is that it is a ground for the change[k] of place by bodies, but the void is not a ground for this, there would not be a separated void; but the void is not a ground for the change[k] of place; so the separate void does not exist. In this hypothetical argument, everyone would accept the hypothetical premiss; he potentially demonstrates the additional premiss[70] as follows: nature is the ground of change[k] of place, but the void is not nature; but he omits as obvious the proposition: therefore the void is not the ground of change[k] of place. It is also possible to support this additional premiss hypothetically: if nature is the ground of the travel of bodies of their own accord, the void would not be the ground of such a change[k]; but the first, so the second.[71] He plainly shows that nature is the ground of change of place from the natural tendency of bodies. For if travel upwards is natural to fire, downwards to earth, and middle places to the intermediates, it is clear that nature and not the void is the ground of change[k] of place. So it seems to some pointless for the void to exist as a ground of this.[72]

664,1 **214b17-19** Again, if there is such a thing as place deprived of body, [when there is void, where will the body go that is placed in it?] clearly not into the whole of it.

He has previously shown that the void is not a ground of change[k] as efficient cause, since even those who posited it did not do so as efficient, but as that in which and as place, in which and through which it was necessary that bodies should move. So by a natural transition he proves that similarly the void as a place in which the moving[k] object moves[k] and rests is neither a ground nor cooperating cause.[73] For if the void is a sort of place, qualified by being void as well as a place by having no body in it, but deprived of that of which it is receptive – if there is, as they say, such a boundless void, where will the body placed in it travel? For what difference in motion[k] will the void give it, being void? It will not be upward or downward exclusively; but if it is simultaneously in all directions either the body will be torn in pieces or it will be at rest rather than in motion[k]. Or rather not even at rest; for why will it rest in this place rather than that?[74]

He used the same argument also in Book 3,[75] when discussing the infinite. For if, he says, there should be a body boundless and homogeneous, its proper place will be boundless, and each part will have the same place, since the whole and its parts have naturally the same place. So, since their proper place is boundless, to what place will the parts of the boundless move[k], or in what will they rest?[76] For they will not be able to move simultaneously everywhere, since they would be torn apart, nor this way rather than that. Similarly they will not be at rest, for why will they rest in this place rather than that? So he produces the same absurdity also for those who say there is a boundless void, there from the likeness of the parts of the boundless to the whole, here from the void being undifferentiated.

For, being undifferentiated, why should what is placed in it move[k] in this direction or in that? So not only will the void not be a ground of motion[k] as that in which it takes place, but, quite the reverse, it will prevent even the natural motion[k] of bodies and, clearly, their natural rest. For, since it is undifferentiated, why should the earth rest here and fire above?

> **214b19-24** The same argument applies also to those who think that place is something separated [into which things travel, how shall what is placed in it travel or rest? The same argument obviously applies to the void also regarding the above and below;] for those who assert the void make it a place.

He first refuted the void through its likeness to the boundless. Now he attacks it from its likeness to the interval of place. For, he says, the argument that will suit those who say that place is a separated interval will also suit those who introduce for us a void that is such 665,1
an interval. For it must be told how each of the natural bodies will travel towards such a place or how they will rest in it. For those who say[77] that place is the limit of the container do not take this limit as being a mathematical surface, but that of a natural body, the special nature of which the limit shares; so, as bodies above differ from those below, their limits will do so as well. Also the natural travel of bodies towards these limits is different, whereas the interval is everywhere undifferentiated. How will above be different from below in it? How will the suitability[78] of fire and water differ in it? It will be the same in the case of the void. For they think that the void is nothing but a separate interval, just like place for those who say that it also is an interval. So if the spatial interval is not a ground of motion[k], clearly neither is the void. It is the same in regard to rest. For why should it be in this part of place or the void or in that? But Alexander says that 'for how shall what is placed in it travel or rest?' would be more consistently taken to mean 'how shall it travel or what is placed in it rest', since being placed is proper to rest. However, surely moving[k] bodies also, if they move[k] in place, could be said to have a position in it. So Aristotle's interpretation seems to be more accurate.

> **214b24-5** and how will things be said to be in a place or in a void?

If this sentence alone follows on the foregoing, there will be no unclarity. For, having said that the body placed in the void will neither move[k] nor be at rest, he reasonably adds: 'how will things be said to be in a place or in a void?' For how in general, when in that in which they can neither move[k] nor be at rest, will they be in it as a place or as a void, whereas both things in motion[k] and those at rest seem to move[k] or be at rest as in a place or in a void.

But in some copies the following is added after the sentence in question: 'For this does not happen when some body[79] is placed whole as in a separate and permanent place; for if a part be so placed, not separate, it will not be in a place but in the whole.' This also would be an argument to show that this separate interval does not exist, whether as a place or as a void. For, he says, how will the body be in such an interval, either as a place or as a void, which comes to repeating 'in a place'? For 'this does not happen', i.e. it does not rationally come about, if place is a permanent three-dimensional nature and some body is supposed to be in it. He adds what comes about irrationally on this view. For it seems that, when some whole body is situated in a place, the whole occupies that place, but its parts
666,1 that are continuous seem to be in that which is in the place, not as such in place; for the parts are then in a place when separated from the whole and their continuity with it is broken. But, while they are continuous with the whole, the place belongs to the whole and the continuous parts to the whole, and they will as such be no less in place than is the whole; for they too similarly occupy an interval equal to themselves. He recounted the absurdities consequent on this in his discussion of place:[80] if the parts are as such in a place, the place will move[k] and a place will be in a place, and there will be many places at once. These absurdities were collected together about place, if place is supposed to be an interval, and he has added to the demonstrations the point that, if such an interval is not even a place, it will not be a void one. For place is supposed to be a void interval, as has been often stated.[81]

> **214b28-215a1** So, for those who say [that void is necessary if there is to be motion[k], the reverse holds, if one examines it, that nothing could move[k], if there were void. For, as for those who say that the earth is at rest because of uniformity, so there must be rest in the void. For no direction is more or less suitable for motion[k].] For where there is void there is no difference.

He has destroyed the arguments that seemed to establish the void, and he has shown that the void is not the ground of motion[k], and therefore is altogether non-existent. Now he proves that quite the opposite of what they intended comes about. For they thought that there could not be change[k] of place if there were no void, but he himself proves that there cannot be absolutely any change[k] of place by bodies, if there be such a void as something separate.

He first, as he had said a little earlier,[82] sets out an argument claiming that since the void is undifferentiated what lies in it will not move[k] this way rather than that. In which case it would not move[k]. He confirmed that, being similarly related to every portion of its environment, it would not move[k] also from Plato, who established the rest of the earth in the middle in this way in what he says in the *Timaeus*:[83] 'a thing set in the middle of the undifferentiated is equipoised.' For if it is similarly related to every part of its environment, why should it approach this part rather than that, since it also has the same tendency towards everything, and the whole surrounding it has on all sides a like relation? Why should it incline one way or another? But, if the void is in every way undifferentiated, and has itself a similar relation to what is placed in it, it too will have a similar relation to the whole in which it lies; so in that way it is equipoised and therefore does not move[k].

I think it worth noting that he seems to disprove natural motion[k] by this argument, if there is a void. 'For there is no direction more or less suitable for motion' accords with this, unless, indeed, it also 667,1 disproves unnatural motion[k]. For unnatural motion is that which is opposite to natural motion. 'It should be mentioned', says Alexander, 'that this argument is not found in some copies, perhaps because it has already been stated, but the one after it is.'[84]

> **215a1-14** A further argument is that all motion[k] is either by force or natural. [But there must be natural motion[k] if there is forced motion[k]; for forced motion[k] is unnatural, while unnatural is posterior to natural. So, if there is not a natural motion[k] for each of the natural bodies, there will be none of the other motions[k]. But how will there be natural motion if there is no differentiation in the void and the boundless? Being boundless, it will have no up nor down nor middle; being void, there will be no difference in up and down. For as there is no differentiation in the non-existent, so with the void; for the void seems to be something non-existent and a privation. But natural motion[k] is differentiated, so there will be natural difference. So either there is no natural motion of anything anywhere,] or, if there is, there will be no void.

He uses this second argument to prove that, if there is a void, there will be no motion[k] – not merely that there will be no natural motion[k], but no forced motion[k] either. So, if all motion[k] is either natural or unnatural, and there can be neither of these if there is a void, it is evident that if there is a void there is no motion[k] whatsoever. He first lays down that all motion[k] is either natural or forced, or, in other words, unnatural, based on the necessity of the division. Next, if there is forced, i.e. unnatural, motion[k], there will certainly be natural motion[k] also, since the unnatural is a diversion of the natural. But universally a diversion is posterior to the normal state,[85] so that, if there were not natural motion[k], there would be none unnatural and simply none at all.

After this prior demonstration, he finally proves that, the void interval being boundless, there would be no natural motion[k], both through it being boundless and through it being void. For natural motions[k] are towards their different proper places. These are above and below, or the middle and the bounds of extension around it. So the boundless will have no above or below, nor a middle and that which naturally surrounds it. Nothing prevents these from existing relationally, but they easily change; but the motions[k] of things that move[k] naturally are always in the same direction. So natural motions[k] could not consist in relational differences, so that if there were no bound there would be no natural motion[k]. The void, being undifferentiated will not permit natural motion[k]; for in it there will be no difference between above and below nor middle and surround. 'For', he says, 'as there is no differentiation in the non-existent, so with the void; for the void seems to be something non-existent and a privation.' So, by conversion, you will truly say that if there is a void there is no natural motion[k]. But also, if there is forced motion[k], there must be natural motion[k] prior to it; but if there is the natural, there is no void, so that even if motion[k] is forced there will be no void. For, if all motion[k] is either forced or natural, it is evident that there would be no motion[k] at all if there were a void. So the required conclusion of the argument in the so-called second indemonstrable also can be derived[86] as follows: if there is a void, there is no natural motion[k]; but there is natural motion[k]; so there is no void. Or, in the first version: if there is natural motion[k] there is no void; but there is natural motion[k]; so there is no void.[87] There can be a fourth formulation as follows: either there is natural motion[k] or there is a void; but there is natural motion[k]; therefore there is no void.

215a14-19 Further, things thrown still move[k] when what impelled them is not touching them, [either through mutual replacement, as some say, or because the air that was pushed pushes them with a faster motion[k] than the travel of the impelled object towards its proper place. But neither of these holds in the void,] nor will travel be possible save as being carried.

Since he has previously disproved forced motion[k], if there is a void, *via* natural motion (since, if there is no natural motion[k] if there is a void, there can be no unnatural motion[k] either) he now provides the direct proof that if there is a void there is no forced motion[k]. Forced motion[k] occurs in two ways, either with the forcing agent being present and carrying, pushing or dragging, or with its not being present, as when things are thrown. From this latter he constructs his proof, omitting the former, as Alexander says, because of its obviousness. For, he says, when the forcing agent is present and causes motion[k] by motion[k], as it moves[k] naturally it necessarily forces the other to do so.[88] For if the forcing agent were itself forced, and that by another, we should go on endlessly. But natural motion[k] is not found in the void, as was proved earlier.[89] But perhaps he did not now omit this type of force because of the obviousness of the proof, but because that type of proof had already been used.

Now the project was to disprove forced motion[k] directly,[90] not *via* natural motion[k]; but he will add the latter a little later. So now he proves that forced motion[k] through throwing in the absence of the forcing agent will not occur in a void. For if something is thrown in what is filled, what is thrown will move[k] either (a) through mutual replacement with the air which is pushed in front of the projectile through the impulsion of the thrower. For the air, being more mobile[k] than the projectile, is pushed in front and, being concentrated by the force, pushes the moving thing on by mutual replacement; as this is continuous, the motion[k] of the projectile remains continuous until, as the impulse of the replacing air gradually weakens, the natural travel of the projectile overcomes it and thus the projectile falls. Plato seems to be of this opinion, expounding mutual replacement as follows in the *Timaeus*: 'as the fire falls out thence again, since it does not go out into a void, the adjacent air pushes the damp mass while still easily moved[k] into the seats of the fire and, compressing it, mingles it with itself.'[91] So the motion[k] of things thrown remains continuous either thus or (b) because the air that has also been pushed away by the thrower and is carried along by the force pushes on the projectile, being more mobile[k] than it so long as it has the power coming from the thrower, and pushes the projectile forward as the air flows on after it, being concentrated through the force of the motion[k], and it adds its pressure until the power in it gradually weakens and the natural travel downwards of the projectile overcomes it. For air is easily mobile and needs some cause for its motion[k] and travel, which it receives in the forcible sending off of the projectile; it is, as it were, sent off together with the projectile and, flows condensed upon it. Doing this with force it moves[k] the dispatched projectile when the thrower is no longer present until its forced and concentrated travel is gradually overcome by the projectile's natural downward tendency.[92]

Thus the motion[k] of projectiles through the filled is explained. But neither of these causes could operate in a void. For nothing could push or mutually replace, since there is nothing in a void. But having said that neither of these is available in the void, nor is travel possible in it, he added 'except as what is carried', thereby showing that it is not possible for what travels in a void to travel through being thrown; if, indeed, it does so when carried. But, if it does, it is carried either by the void, or by some body. But this is impossible in the case of the void, for the void is impotent, being nothing. But if there be some body to carry things in the void and move[k] them forcibly, since it moves[k] being itself moved[k], it is evident that it is moved[k] either naturally (but it was proved that natural motion[k] in a void was impossible), or by force. If by force then either (impossibly) as a projectile, or again as carried, and so on without end. But since, as has been said, forced motion[k] is (1) by throwing or (2) by the presence of a moving[k] force, and the latter forces either by pushing or pulling or carrying, he took carrying as representing all, carrying being more general since pushing and pulling are a sort of carrying.

But what prevents animals from moving[k] themselves by their own effort in a void and from moving[k] other bodies forcibly, by either pushing, pulling, carrying or even throwing? For one cannot say in the case of animals that they will move[k] naturally upwards or downwards exclusively, with the result that even animals cannot move[k], since up and down are absent in a void. But since our subject is natural science and we are now discussing natural, not intentional,
670,1 motion[k], he will discuss the latter in *On the Movement of Animals*.[93] But as far as things thrown are concerned, even if they be thrown by animals in the absence of another mover, things moving[k] contrary to their own nature are a topic of natural science.

> **215a19-22** Further, nobody could say why something moved[k] should come to rest [anywhere. Why here rather than there? So it will either remain at rest or inevitably travel on endlessly,] unless something more powerful stops it.

Having previously proved that if there is a void there will be no motion[k], either natural or unnatural, I think that he now proves that, if there is a void, not only is motion[k] abolished, but also natural rest. For the natural rest of bodies in their own places comes about through differences, while a void is undifferentiated. So why should something moved come to rest here rather than there, towards which place and because of which bodies move[k]? So, if there is no difference in the void, even if one posits bodies moving[k] in it, they will inevitably move[k] endlessly unless some stronger body stops them. And if they be said to be at rest, they will rest for ever. For if it is natural for them to rest in this or in that part of the void, they will clearly do so in all. So it would be proved that it is not even possible to be naturally stationary if there is a void. For 'or inevitably travel on endlessly' is added as consequent on not being naturally stationary.

If the commentators[94] call this also an argument that there will be no natural motion[k] in a void because things moving[k] naturally travel towards their own place, and there is no place of their own in a void, since it is undifferentiated, I think that it is better that, since motion[k] has been previously abolished, it is rest that is now abolished. Also this seems to harmonise better with the text which begins the argument as follows: 'further nobody could say why anything moved[k] should come to rest anywhere.' But he omitted to enquire what could there be in a void such as to prevent motion[k] by being more powerful, since it is evident that there is no such thing. For that also would prevent while being either naturally or unnaturally stationary. It cannot do so naturally and, if it were itself prevented, there would be an infinite regress. But even if someone should say that something moving[k] should force the travelling body to stop, this absurdity has already been discussed.[95]

> **215a22-4** Further, things are now thought to move into a void [because it yields. But in a void it is thus throughout,] so that they will move in all directions.

He attacks from all sides the notions on account of which they posit the void. So now, since those who assume a void because of its yielding nature and say that the cause of motion[k] through air and water but not through stone and earth is that air and water contain many voids but stones and earth are more full, which is why they do not yield, what, he says, will they say about the boundless void? For, since it is equally yielding in all directions, it will contain no difference for bodies favouring motion[k] into this part or that. So they will either move[k] in all directions or nowhere.

Alexander says that 'it is possible to use this argument also against the Stoics who say that a boundless void surrounds the universe. For, if the void is boundless, why does the universe stay where it is and not move? Or, if it does move, why rather here than somewhere else? For the void is undifferentiated and equally yielding everywhere. But, if they say that it rests as being held together by its own holding power, its holding power might perhaps contribute to its parts not being scattered, torn apart and dispersed in all directions, but nevertheless the holding power will contribute nothing further towards the whole remaining with its holding power and not travelling away. It could', he says, 'be thus proved that the void is not the cause of motion[k] even as the medium, if what is placed in it has no cause to move[k] in any direction. But it is clear that he drew this conclusion also from earlier arguments.'[96]

> **215a24-b21** Our statement is clear also on the following grounds. [We see a body of the same weight moving faster than another[97] for two reasons, either because of a difference in the medium, as through water or earth, or through water or air, or else because of a difference in what travels, if, other things being equal, there is excess of weight or lightness. The difference in the medium is a cause because it obstructs, especially if it is travelling in the opposite direction, but also if at rest. What is not easily divided, which is the more dense, does so more. The body *A* will travel through *B* in time *C*, but through *D*, which is thinner, in time *E*, if *B* and *D* are equal in length, according to the ratio of the densities of the obstructing bodies. For let *B* be water and *D* air; *A* will move through *D* faster than through *B* proportionally as air is thinner and has less body than water. The ratio of speed to speed will be the same as the difference between air and water. So, if it is twice as thin, it will traverse *B* in twice the speed in which it traverses *D*, and the time *C* will be twice the time *E*. Always motion will be faster proportionally as the medium has less body, is less obstructive and more easily divided.
>
> But there is no ratio by which the void is exceeded by body, as zero has no ratio to number. For, if four exceeds three by one, and two by more, and one by even more than two, there is no longer any ratio by which it exceeds nothing. For that which exceeds must divide into the excess and what is exceeded, so four will be both the excess and nothing. That is why a line does not exceed a point, unless it is composed of points. Similarly the void can bear no ratio to the full,] so neither can their motion[k].

He now adds to the argument the most complicated demonstrations, proving that there will be no motion[k] in the void. There are four arguments of this type that end in a reduction to the impossible. The first concludes that motion[k] will be timeless, the second that the void will stand in a ratio to the full, the third that the void will not only stand in a ratio to the full, but also in that which the full has to the full, the fourth that the moving[k] object will traverse the void and the full in the same time. He adds the causes of the different speed of motion[k], two in number; for if the moving[k] objects are of the same species (e.g. both naturally moving[k] downwards or both upwards) and have the same shape, the difference in speed results either through a difference in the medium (e.g. through earth or mud or water or air, and if one were to move[k] through a stationary medium, the other through one in opposite motion and colliding with it), or else one moves[k] faster and the other slower through the same interval, either when the moving[k] objects have the same natural impulse (e.g. upwards or downwards) and the same shape, but one has an excess of weight, when both are heavy, or of lightness, if both are light. For, other things being equal, the heavier will move[k] faster, as if a ball of gold and one of silver were both to move[k] through air.

But this difference will be explained a little later.[98] Now, assuming that the media of the motion[k] are one with denser parts, the other 672,1 with lighter parts, and seeing the difference of speed as occurring in these conditions, he adduces the following absurdities against those who say that motion[k] is through a void.

Let a constant weight move[k], if you like, in one hour through the air that is thinnest a distance of two hundred yards. But now how long will it take it to move[k] through two hundred yards of void? If, perchance, in half the time, the air will be twice as dense as the void. If in a third or a tenth or a thousandth of the time, the density of the medium will be in the same ratio as the time is to the time. But one can find no ratio between the full and the void, for it will exceed any suggested. The result is that the same weight will move[k] in the void in no time at all through the same distance as it moved through the full in a given time. So there will be timeless motion[k], which is impossible, if all motion[k] occurs in time. He proves that the void stands in no ratio to the full through the similar relation of nothing to number. For nothing stands in no ratio to number, by which it is exceeded; nor would number be said to exceed nothing.

He proves this also by positing an excess between things and thus provides a criterion of excess. For, since everything that exceeds is divided into that which is exceeded and the excess, it is clearly composed of these. For a thing is composed of the constituents into which it is divided. If we take the excess of the number four in relation to nothing, it is clearly the same four. Therefore the number four is composed of four and nothing. But it is absurd to call nothing a part of the number four, and to say that it is composed of nothing. This, he says, is why a line does not exceed a point, because it is not composed of a point, nor is a point a part of it.[99] For what is exceeded must be in what exceeds as a part, and what exceeds cannot be the excess on its own. So, as a number does not exceed nothing, it also stands in no ratio to it, just as a body has not to the void, so that a body should not be composed of body and void. Consequently, neither the movements[k] through them nor the times in which the movements occur will be in any ratio. For motion[k] through a void does not occur in time. As is customary, for exactness he makes his exposition with the use of letters: he calls the mass that moves through the different bodies *A*, the denser medium he calls *B*, the thinner *D*, the time in which which the determinate magnitude *A* moves[k] through *B* he calls *C*. He says that clearly *A* will move[k] through *D*, which is thinner than *B*, an interval equal to that which it moved through *B* clearly in a lesser time; the time will be less proportionally to the greater density of *B* which impedes the motion[k] of *A* more than *D*, which is thinner and impedes less. Wishing to make the proportion clearer, he names the bodies through which *A* moved, and calls the denser one, *B*, water and the thinner one, *D*, air. He says that proportionally as air, which is thinner than water and has less body, i.e. impedes less, so *A* will move faster through it than through water.

In this way he concluded the first impossibility, that there will be timeless motion[k].[100] For if void stands in no ratio to the full, but motion[k] is related to motion[k] as void is to the full through which the motion[k] occurs, and if times measure the motions[k], clearly one time will be to the other as nothing to number. So there will be no time. For every limited time stands to every other limited time in some ratio, as every number to every number, so that there will be a motion[k] – that through the void – which is not in time; or, if it is in time, some motion[k] will be related to some limited motion[k] and some time to some limited time in no ratio, which is impossible.

Aristotle seems to me to exhibit this impossibility best at the end of this passage, where he says: 'Similarly the void can bear no ratio to the full, so neither can their motion[k].' It will be clear from what follows that he takes time together with motion[k].[101]

> **215b21-6** But if it moves such a distance in such a time, through the thinnest medium, [it will exceed every ratio through the void. For let *F* be a void equal to *B* and *D*. If *A* is to traverse it and move in a certain time *G*, that is shorter than *E*,] the full will have this ratio to the void.

He has proved by the first argument that the void is in no ratio to the full, and concluded from this that, if there is motion[k] through the void, the consequence for motion[k] and time is impossible, since a determinate motion[k] will be in no ratio to a determinate motion[k] nor a determinate time to a determinate time; for this shows that 'so neither can their motion[k]'. Another consequence is that there will be motion[k] that is not in time.

Now, having said that motion[k] through the void will be in no ratio even to that through the thinnest of media, since motion[k] through the void will be timeless, he adds what follows for those who do not make these impossible claims. Alexander treats 'But if it travels such a distance in such a time through the thinnest medium, it will exceed every ratio through the void'[102] as the conclusion of the previous argument.[103] He began the one now projected from 'for let F be a void'.

But I think it better not to treat this as a beginning; otherwise what reason will there be for the conjunction 'for'? By the former he shows what follows for those who posit that motion[k] through the void is in time and say that there is a ratio between the motions[k] and the times. The inverse follows that there will be a ratio between the void and 674,1
the full, which he proved to be impossible. For, just as it follows from there being no ratio between the void and the full that there is one neither of motions[k] nor of times, so it follows from there being a ratio of *both motions[k] and times that there is a ratio*[104] of the void to the full.

For he has posited that motion[k] through a void occurs in a determinate time, and assumes the same proportion in the times taken by the motions[k] and their media. From these he draws an impossible conclusion, that there will be the same ratio between the void and the full as time to time, or, in other words, number to number. And not this alone, but also that the void will have the ratio to the full which the full has to the full. For, he says, let the void *F* be equal in size to *B*, which was water, and to *D* which was air. But if the moving[k] object *A* moves[k] in a certain time, say *G*, through the void *F*, the time *G* will clearly be a time less than *E*, in which *A* moved[k] through *D*, which was air. For this was the time stipulated for motion through this medium. Also the void will have to the full the same ratio as the time *F* to *E*. For the motions[k] and the times taken by the motions[k] will be in the same proportions as the substrates through which the motions[k] occur are to each other. Conversely, as the one time to the other, so substrate to substrate. So, if there is a ratio between every determinate time and every other determinate time, there will be a ratio of the void to the full. This is why he added that 'the full will have this ratio to the void'; given that it is impossible for a number to be in a ratio to nothing, as he proved earlier.

> **215b26-7** But in the quantity of time *G* *A* will traverse the part *H* of *D*.

Based on the same posits, he concludes a third impossibility, that the moving[k] object will traverse full and void in the same time. Thus *A* will also move[k] through something full in time *G*, in which it moved[k] through the void *F*. For, if in time *E* it has traversed the whole of *D*, *A* will move through a portion of *D* in *G* that is less than *E*. *G* is less than *E* since in *G* it moved[k] through the void *F* a distance equal to *D*. Call this portion *H*. The conclusion is clearly brief, but obvious, that the moving[k] object *A* will traverse the void *F* and the full *H* in the 675,1
same time *G*, though they are not equal in length. Alexander says: 'Having brought the discussion so far, he neglected to demonstrate the consequent absurdity. This is that the void will have the same

ratio to the full as the full to the full. For, as the time *G* is related to time *E*, so the part *H* of *D* will be related to the whole of *D*. But now, the relation that the time *G* had to *E* is the same as that between the void *F* and the full *D*, but things having the same relation to the same thing have the same relation to each other; so the ratio that the body *H* has to *D* is the same as that of the void *F* to the full *D*. Therefore the void has to the full the same ratio as the full has to the full.' So Alexander. And perhaps the absurdity of which he speaks also follows; but Aristotle concluded the traverse of void and full in equal times as obviously absurd and impossible. He says later[105] in his discussion that 'this absurdity will come about. It will be found to traverse what is full and a void in the same time.' For one body will be in a proportion to another as time to time, to state a conclusion recently drawn. For because the time *G* is the same part of the time *E* as the magnitude *H* of the magnitude *D*, therefore the time *G* is the same in which *A* will traverse the void *F* and the full *H*.

If someone were to think that nothing absurd was being concluded, because the interval of the void *F* is greater than the interval of the body *H*, let him know that Aristotle did not wish to conclude their equality, but a similarity of ratio. For since he proved earlier[106] that the void has no ratio to the full, just as nought has none to number, he now reasonably concluded as absurd the traverse of full and void in the same time.

215b27-216a12[107] It will surely traverse anything there may be thinner [than the air *Z* in the same proportion as the time *E* has to time *G*. For if *F* be thinner than *D* by as much as *E* exceeds *G*, *A* will, if it travels, traverse *F* in a time equal to *G* inversely proportional to the speed of the movement. But if there be no body in the interval *F*, even quicker. But it was to be in *G*. So it will traverse the full and the void in the same time. But that is impossible. However, it is manifest that if there is a time in which it will travel through some void, this impossibility will come about: it will be found to traverse what is full and a void in the same time. To sum up, the cause of what happens is clear: there is a ratio of every motion[k] to every other motion[k], since they are in time and] there is a ratio of every time to every other time, since both are limited. But there is none between the void and the full.[108]

Alexander says: 'In this passage he adds that even if some body exceeds in thinness body *D* by as much as the time *E* exceeds *G*, *A*
676,1 will traverse in time *G* a part of the thin body equal to what it does of *D* that is more dense in *E*. He now posits *F*, which was formerly the void, to be thinner than *D*. For if the body *F* be thinner than *D* as

the time *E* exceeds *G*, the body exceeding in thinness will traverse in inverse ratio the exceeded time, but the body exceeded in thinness in the exceeding time – thus the denser in the greater, the thinner in the lesser.' Alexander explained well the content of the passage, but he did not say of what use it is to the task in hand.

However, I do not think that this discussion is a mere addendum. But he says that it is not even probative; for it will not always be possible to find something thinner than *D* by the same amount as time *E* exceeds *G*. In the first place, there may be nothing thinner than *D*; and it is possible for there to be something thinner than it, but not by the amount that time *E* exceeds *G*. 'To think', he says, 'that all said before, that was so strong, was leading up to this, and was said for the sake of this, that is so flawed, as the commentators say,[109] is a mark of those who have not followed the proofs nor comprehended that this is added *ex hypothesi* and not as probative. That is why he used the words "will traverse anything that may be thinner", indicating thereby that there is no necessity but that, if it be so, the same proportion will be preserved.'

It was reasonable to say that there will not always be something thinner than *D*. But perhaps Aristotle, in positing that *D* was air and that *H* was a part of *D* and taking *F* to be void, wished to extend the argument to the very thinnest and nearest to the void through its thinness, in order to exhibit clearly the proportion and, by the gradual diminution of the time, excite the idea that motion[k] through the void, if it existed, would not be in time. For if the traverse of the void had to be faster than the least time in which the thinnest medium is traversed, it would clearly be no time at all. That, I think, is why he added 'if it travels' in the case of *F*, because, if it were thin, *A* would traverse it in time *E*; but, if it is void, it will not move[k]. Further, if it were posited generally that it would move[k] through *F* that is void, the envisaged impossibility follows more plainly. For it will move in a lesser time. But the motion[k] of *A* through *F* that was void was to be in *G*, in which time it was proved that it would traverse body *H*. 'So it will traverse something full and a void in the same time, but that is impossible.' He adds the cause of the impossibility. For if something moves[k] through the void in some time, it will result that a thing traverses the full and the void in the same time, because, as was also said earlier,[110] as *D* is to *H*, so is *E* to *G*. But *D* was to be to *F* as *E* to 677,1
G. Therefore *D* will be to *F* as *D* is to *H*. So *F*, which is void, will have a ratio to *D*, which is full, which the full *H* has to the full *D*. A thing will traverse the void and the full in the same time, which is absurd, even if they are unequal. He himself showed that all these absurdities followed from saying that motion[k] through the void occurred in some time, when he said, 'It is however clear that if there is a time in which

it will travel through a part of the void, that is impossible',[111] i.e. if something were posited to move[k] through the void in time.

Next, he well added the reason why these impossibilities follow from the hypothesis that says that there is motion[k] through the void. For, if there is some ratio between motion[k] and motion[k], and time and time, both being determinate, but none between void and full, it is clear that a motion[k] through a void will be timeless. But if someone were to posit motion[k] through the void to be in time, then, since time is in a ratio to time and since the substrates through which the motion[k] occurs are as the times taken by the motion[k], the void will also be in the ratio to the full in which the full is to the full, and a thing will move through the void and through the full in the same time.

These, then, are the absurdities that are consequent on the difference between the media through which things travel, apart from their consistency and this their cause. It is easy also to abolish the void by using the other cause that arises from contrary motion[k].[112] For the void will not be in contrary motion[k], since it is not in motion[k] at all.

But, perhaps, if one were to posit that the void was something separate and were to say that motion[k] was through the void (since it was not through the full), but that the void always contains some body, and that motion[k] through the void occurred as this body was thrust away by the moving[k] object, he would not be entrapped in any of the absurdities already alleged nor those to be alleged. For let it be conceded that, so far as the void is concerned, bodies move[k] at the same speed; but the difference comes about through the bodies in the void, which are pushed aside more quickly or more slowly, and the bodies travelling must behave similarly. So, if one body were dense and hard to divide or hard to penetrate, another the reverse, the speed of motion[k] becomes unequal, even though it is all through the void. The difference also comes, as he himself has said and will say, from the moving[k] objects themselves, if some have a greater impulse and some less. As from other powers, bodies have also differences through powers arising from their impulse. In this way motion[k] will not be timeless, nor will the void be in a ratio to the full. For motion[k] is never through a void alone, nor does a thing travelling do so more quickly only through the medium being more easily divided, but also because of its own powers.[113]

That motion[k] is not through a void interval can be established as follows, as has been said in the discussion of place.[114] The moving[k] body moves[k] from somewhere to somewhere; the interval between the places whence and whither the motion[k] is either void or an interval consisting of a body in between, e.g. air. But the interval of air is conjoined with the air. How, then, will a moving [k] object traverse an interval that is nothing? And how will it be measured?

216a12-21 The differences because of excess in the things travelling are these: [we see things that have a greater impulse either of weight or of lightness, other things being equal, travelling faster an equal distance in the ratio that their magnitudes have to each other. So through the void also. But this is impossible. For what reason will they travel faster? They will necessarily do so in full areas, for the greater divides the medium more quickly through its strength. For what travels or is thrown divides either through its shape or through its impulse. So everything] will move at the same speed; but this is impossible.

Two causes of unequal speed of motion[k] have been stated. One is the difference of the media through which the motion[k] takes place, if one be more dense and thick, the other thinner and yielding; or if one is static, the other in opposing travel. The other is the difference between the moving[k] things themselves, there being several causes in these of unequal speed. There is shape and impulse, which is different either because of their consistency or their size. For gold that is smaller is heavier than a bigger piece of wood and a bigger clod than a smaller. Keeping other things equal, but changing the impulse in the substrate bodies, he proves in this way also the resulting absurdities for those who say that motion[k] occurs through a void. For, when other things are equal, we see things with a greater impulse of weight or lightness travelling more quickly downwards and upwards. Thus if two balls, one of gold and one of silver, of equal size be released, the golden one will fall faster.[115] This would have to occur similarly in the void also, if motion[k] were to occur through a void. But, indeed, that will not happen in a void. For what will be the cause in this of one travelling faster, the other slowly? In the full, the heavier object will divide the substrate more quickly, but in the void there is no place for such an explanation; for there is nothing to divide. So moving[k] objects will travel at the same speed, even if there is a difference in impulse. But this is impossible; for to be heavier is essentially to travel faster downwards.

Alexander writes 'but this is impossible at the same time' instead of 'but this is impossible'[116] and explains: either because it is impossible for a lighter and a heavier body to travel downwards at the same time, or because it results that the heavier object, through its own nature, must necessarily move[k] faster, but will not move faster through the void. But these are at the same time impossible. As to 'in the ratio which the magnitudes have to each other', he says that the travelling bodies travel according to the ratio which they have to each 679,1
other in impulse. They travel faster or slower by so much; e.g. if one

is twice as heavy, it will travel downwards twice as fast, i.e. in half the time.[117]

Also, those who introduce the void do not provide an explanation of the difference resulting from shape. One wants to know why a wide piece of iron or lead is more inclined to rest on the surface in water, but a rounded or long one is not, even if it happens to be much smaller. It is easy for others to say that the broad things do not move quickly downwards because they cover much air or water, on which they ride; for the substrate is not easily divided because of the bluntness of the shape. But things shaped otherwise cover little, which is why they rend very fast the substrate in the direction of their impulse. But the advocates of the void cannot give this explanation. 'On this basis', says Alexander, 'it is possible to say to Epicurus, and perhaps equally to Democritus and Leucippus, and in general to those who make the principles to be the atoms and the void, that if atoms travel at unequal speeds through the void, it is time that they stated the causes of the difference in speed: since for them[118] size or weight or shape contributes nothing to the speed. But if they move at the same speed, one will never catch up with another, nor will they hit into or enclose each other. For the difference in their shapes cannot make their speed different. For shapes make speed of travel unequal by dividing or not dividing, but in the void nothing is divided. So, according to them, there will be no coming to be.'[119] Perhaps this absurdity would follow for them if they were to say that the the atoms are in everlasting motion. For if they both rest and move[k] at equal speeds they will catch up on each other. But if one were to enquire why it is impossible for a heavy and less heavy object to move[k] at equal speeds, it is easy to say that if an increase in weight does not make the motion[k] faster weight will not in the first place be a cause of downward motion[k] in the void. For what, when added, does not intensify or increase motion[k] cannot at the start be a cause of motion[k]. But if impulses are not the cause of motion[k] for bodies in the void, bodies would not move[k] at the start in the void. For there is no other explanation of the natural travel of bodies save their innate impulse.[120] Alexander adds: 'If it is thought by the Epicureans that every body has weight and that bodies move[k] through the void because of their weight, it follows that for those for whom weight is the cause of motion[k] a greater impulse will be the cause of faster motion[k]; but this cannot be so in a void, nor would they move[k] in it through their weight. But, if not through this, they would not move[k] at all.'

Hearing this, I think that one would at once say that if impulse is the cause of motion[k] for bodies, and the greater [the impulse] the
680,1 faster [the motion], why will bodies not move[k] in the void? For even if ease or difficulty of division contributes something, as the medium through which the motion[k] occurs, to the greater or lesser speed of

the motion[k], still this is not the cause of the impulse. But, if there is this difference, the ease or difficulty of division adds something to the difference in the natural impulse. For, just as other qualities of the same kind are greater in greater things, so it is with impulse. In general, the impulse is the cause of the division, not the division of the impulse, whether the substrate obstructs, or not.[121]

> **216a21-6** From what has been said it is clear that, if there is a void, the opposite comes about [from what those who say there is a void depict. For these, on the one hand, think that there is void if there is to be change[k] of place, divided off on its own. But this is the same as to say that place is something separate.] But this was earlier said to be impossible.

Those who introduce the void did so as necessary for motion[k], since it was impossible to move through the full. So they proposed an hypothetical argument saying 'If there is motion[k] there is void.'[122] But Aristotle proved in many ways the opposite, saying that if there is void there is no motion[k]. He also recalls the proof stated in the discussions about place that place is not a separate empty interval.[123] For the argument is apposite now. It is the one that concludes that place will move and that there will be many places together.

Having said: 'For these, on the one hand, think that there is void', he did not immediately provide the 'on the other hand', but first replies to these people. He adds the following about those who posit the void in other ways: 'But there are some who through the rare and the dense ... .'[124]

> **216a26-b12** Even considered on its own the so-called void might [seem truly void. For, just as if one immerses a cube in water, a quantity of water equal to the cube will be displaced, similarly in the air, though imperceptibly. Indeed always any body that permits change of position will necessarily do so, unless compressed, either always down, if it travels downwards, as in the case of earth, or upwards, if it is fire, or in both directions, depending on what it is immersed in. But this is impossible in the void, since there is no body. An equal interval must pass through the cube as there was earlier in the void, just as if the water did not yield position to the wooden cube, nor air, but penetrated it throughout. But, now, the cube has a magnitude as great as that occupied by the void. If this is also hot or cold, heavy or light, it is none the less different in essence from all its states, even if not separate from them. I refer to the mass of the wooden cube. So if it were separate from all the rest and were

neither light nor heavy, it would occupy an equal void and be in that which is the part of place and the void equal to itself. How then will the body of the cube be different from the equal void and place? And if there may be two such, why cannot there be any number in the same place?] This, then, is one absurdity and impossibility.

The teaching is highly organised. For first he dissolved the arguments that seem to establish the void, and then proved that it is not the cause of motion[k], and therefore does not exist at all, if it was introduced for this. He then turned the argument to the opposite conclusion. For while they said 'if there is motion[k] there is void', he proves by many arguments that if there is void there is not motion[k]. Finally he sets out the problem in itself, examines it, and proves that there is no void, reducing, in turn, to impossiblity the view that body would pass through body if there were no[125] void, to avoid which they posited the void.

For when one body is placed in another, e.g. a stone cube in water or in air, that which was there before must necessarily make way for
681,1 what enters, upwards if it is fire, or downwards if it is earth, or to the middle, if it be one of the intermediate elements, and make way in proportion to the bulk placed in it. This is clear in some cases, such as of a stone cube being thrown into water. The same bulk of water as of the cube will make way.

In certain cases the yielding of place is not very obvious to perception, but is understood from certain apparatus. When water is poured into hydraulic organs that are full of air, if reeds of trumpets[126] or shawms are attached to the holes through which the air stream comes out, they prove the exit of air through them by the sound. But, if the previously contained body has no exit and another is forced in, either the contained body is compressed into itself and becomes dense as it is pressed together in order for the container to receive a quantity of added body equal to the compression of the first volume, or the container would burst rather than receive more body.

When this is all so evident, what shall we say happens in the case of the void, when a body is placed in it? That an equal quantity of void yields its place? But that is laughable, if the impossible is laughable. For what is it that yields its place? That it remains? Then an equal interval of void will permeate the cube; so that, if the void is altogether nothing, we shall have what we are seeking. But, if it has some nature in three dimensions, how will it permeate another similar interval? That is equally as absurd as if the water did not make way but were to pass into the stone cube. Why are they prevented from passing into each other and occupying the same place, but not the void? Is it because they are warm, or white, or heavy, or filled with some other features they happen to have, but the void has no such features? Or is this an absurd supposition? For it was said earlier as well that bodies have a place in virtue of those intervals alone that they need, even if scattered about, for the position of their parts.[127] For, even if the intervals are inseparable from other attributes, their essence as intervals, which their definition naturally separates off, is something different. The stone also does not occupy so much place because it is white, or black, or heavy, but because it occupies an interval of that size. However, it travels into this place because of its weight, but, when it is in a place, it is in the place because of its volume. For the size of the place follows from this alone among the attributes of the cube. So, if one posits a stone cube separated from all its other attributes, but just being in an interval, it will none the less occupy the space that it did when it had its other features. So how will the interval of the cube differ from the equal void and place? Also, if two such things can be in the same place, why not more and even an infinite number?[128] Drawing the argument together, you will see the necessity of the demonstration. For, if there is a body in a void, an interval is in an interval and a volume in a volume; in that way the body is in a place. But if there is an interval in an interval and a volume in a volume, there will also be a body in a body. For the other attributes in no way prevent the bodies from permeating each other. So, if it is impossible for a body to be in a body, it is also impossible for a body to be in a void. He called it truly void, because it would be shown to fill no useful purpose.

216b12-21 Also it is clear that the cube will remain, [even when displaced, in the same condition as other bodies. So if it is in no way different from its place, why need one create a place for bodies beyond their volume, if the volume is unaffected? For nothing is gained] if another equal interval of that sort is attached to it. So it is clear from these considerations that there is no separate void.

He sets out this second argument with the same purport. For the interval in each thing, separated in definition from its conditions, i.e. its attributes,[129] is no different from the void, and, when moving[k] and transferred, bodies move[k] with their own intervals, what need have bodies of other such intervals? For if, as each has an interval, it needs another interval, why will not the void also need another, and that another, and so *ad infinitum*?

216b22-30 But there are those who think that it is clear that there is a void because of rarity and density. [For if there is not rarity and density it will not be possible for there to be contraction and condensation. But, without these, either there will be no motion[k], or the universe will bulge, as Xouthus said, or air and water must always change equally (I mean e.g. that if air came from a cup of water, at once an equal quantity of water will have come from air), or else there must necessarily be void. For compression and dilation] is not otherwise possible.

Of those who assert the void, some said the void was something separate existing on its own, spread through all bodies, and in parts of itself having bodies and in other parts not, which they also called the true void, as having a place. Others said that it was scattered around in small quantities everywhere in bodies.

Having addressed the former view he now sets out the other. In the former he said, 'so some think that there is a void, if there is change[k] of place, separate and on its own'.[130] The present one he introduces by, 'But there are those who think that it is clear that there is void because of rarity and density.' For these said that if there is
683,1 rarefaction and condensation, there is void. He has already spoken against this argument. For this was one of the four arguments at the beginning that posited the void to be refuted by the truth;[131] of these some introduced the separate and self-contained void on account of motion[k], others the interspersed void on account of rarefaction and condensation. Now he sets out still more precisely the structure of the argument and adds its destruction.

These people said not only that the elimination of the void involved the elimination of rarefaction and condensation, but that if there is no void there will be no condensation and rarefaction and that without rarefaction and condensation there is no change[k] whatever; neither change[k] of place, nor growth, nor alteration and coming to be. For they say that change[k] of place comes about only by the contraction and compression of bodies, thus making room for things that move through them, as if for walkers through a crowd. Also things that grow and spread out in bulk do so as other things grow more dense and contract into the voids in them, and thus find space for increase. But also things that become bigger from smaller and take up a bigger space do so by some being compressed and contracting. But compression cannot occur unless void is interspersed in bodies, nor can anything become bigger from smaller in any way unless there is rarefaction and there is no rarefaction without void separating the bodies. Also there would be no alteration without change[k] of place. For what causes and what suffers alteration must come together by motion[k].

So they said that if there is no void, there will be no rarefaction and condensation; but if there is no rarefaction and condensation there is no change[k]. But, if there were change[k] without rarefaction and condensation, the universe would bulge, as Xouthos the Pythagorean says, and it will heap up and spread further as the sea heaps up on the beach with its waves.[132] This comes about[133] as things changing[k] their place in a straight line push forward things in their way, as do also things growing, since there is no compression by which space can be made for approaching things. It is only in the case of things in circular motion[k] that exchange of place occurs; for in this case only did they allow reciprocal exchange. When one of the densified bodies becomes more rare it must push out again and pour over. But the pouring over would not occur in its turn, unless there was void, clearly that which was separate and outside the universe, so that the argument returns to the same point: the void being eliminated, compression is eliminated, while its elimination introduces the void. For the universe cannot bulge into body, but always into a void.[134]

There is also another artificial hypothesis concerning things coming to be. For someone might say that when a cup of water is transformed into more air, other air somewhere equal to it preserves equality by being transformed into a cup of water. Or fire, also, by sometimes changing into air guards the ratio. That, I think, is a better understanding of the text than Themistius' understanding that 'it is necessary that a cup of air comes from a cup of water'; that is impossible.[135] Perhaps that is also Alexander's interpretation when he says that the transformation is of equals into equals; unless he says that each pair is equal to each, and not the air to the water, as does Themistius.

So if there being no change[k] and the overflow of the universe and the equality of the transformations are all absurd, void must necessarily be mixed with bodies. Only thus will there be compression and rarefaction, and none of the above absurdities will follow.

216b30-217a10 So if they call the rare [that which contains many separate voids, it is clear that if there cannot be a separate void, nor a place having its own interval, there is also nothing rare. But, if the void is not separate, but is none the less inside, it is less impossible, but it results, first, that the void is not the cause of all motion[k], but of that which is upwards; for the rare is light, which is why they say that fire is rare. Further, the void is not the cause of motion[k] as the medium, but like skins which, as they travel up, carry with them what is continuous with them, so the void carries things upwards. But how can there be travel by a void or a place of a void? Its destination will be void of void. Also, what do they say about the heavy thing that travels downwards? And it is clear that if a thing travels upwards in proportion as it is rarer and more void, it would travel fastest if it were wholly void. But perhaps it is impossible that it should move[k] at all. The argument is the same, that just as everything cannot move[k] in the void, so the void itself is motionless[k].] For the speeds are incommensurable.

Having supported so far as possible those arguments that introduce the void because of the rare and the dense, he sets out to attack them, and again divides the void into the separate and the inseparable. He calls a separate void an interval as such that is continuous and receptive of body but not actually occupied; he calls unseparated that which is interpersed in bodies. For these were the two hypotheses of the proponents of the void. Now, I think, he divides the interpersed into that which is interpersed in large portions and that which is not. He says that those who speak of the rare, if by that they mean that which has many separate voids, i.e. with noticeable boundaries, it has already been shown that it is impossible for there to be such a void which would allow motion[k] through it.[136] The same thing was proved also when we were proving that place was not a separate interval. In general, there would be nothing rare in that way.

If they were to describe the rare as where the void was interpersed in small bits and not separate, the hypothesis is more plausible and seems less impossible, but even now the following absurdities will arise. First, that such a void will not be the cause of all motion[k], but, if at all, of upward motion[k]. For the rare, which is thought to be such, is light and leading upwards because of such a void in it. That is why they themselves say that fire is rare. What then will be the cause of downward motion[k], as he will shortly object? But now, having conceded that the rare is the cause of upward motion[k], he shows the way in which it comes to be the cause of local motion[k], that it is not as the medium through which, as those said who introduced the separate void, but as itself causing motion[k] and being its efficient cause,[137] as nature was shown to cause motion[k]. Appositely and appealingly he added an illustration of blown-up skins in water which also make
685,1 their burdens[138] light, like nets in the sea and their contents. At any rate, people cross rivers using skins to ferry them across.[139] That is the way in which the void seems to make bodies in which they are light, by itself travelling upwards.

If the void is like that, he now points out two absurdities, one being that there seems to be some motion[k] and travel by the void. How can a void travel if it is void? Second, if a void naturally travels upwards, it will manifestly be in the region above as a place, so that manifestly the place will be void, so that the void will be in the void. In addition he brings forward the following absurdity for those who say that the void travels upwards. For let this be the cause of upward motion[k]: what will they say about the heavy, and what cause will they give of travel downwards? For the void will not be the cause of this, since it will not drag down the bodies in which it is. So it is not the void that is for them absolutely the cause of change[k] of place, but something else that is the cause of both downward and upward motion[k]. For as, according to those for whom nature is the cause of bodily motion[k],[140] it is not said to be so of upward motion[k] alone, but of both, so the void, if it were the cause of motion[k], would be said to be so of both the upward and the downward. But if someone were to say that heavy things travel downwards through the natural downward tendency in bodies, which is called heaviness, why do not light things travel upwards through the lightness in them, but nature is lame in not providing an opposite of heaviness? This position fits in with those that say that all bodies are naturally heavy and downward moving;[141] some invoke the void and some the settling of the heavier to explain some things being above others.

He adds another absurdity for those who account for the upward travel of bodies by the void. For, if it is the cause of travel upwards as carrying bodies upwards with it, it is clear that the void will travel up quickest on its own, since it does so when it is carrying bodies. What is the absurdity? one may ask. Is it that motion[k] by the void is totally impossible? He says that the same argument applies as that which earlier proved that there is no motion[k] in a void, since the void stands in no ratio to the full; in that way he proves that the void will not itself move[k]. For if one body moves[k] faster than another in the same medium through its excess, as of weight in things travelling downwards, and so, in the case of things travelling upwards, of lightness and rarity, and proportionally in less time the lighter they are, since the light body stands in no ratio to the void and does not exceed it, there would be no time in which the void moves upwards with the same motion as the light body. For the void exceeds everything bodily in lightness by an incommensurable excess. For void will 686,1
have to body the proportion that the time in which the void moves has to that in which the light body does so. But void has no proportion to body, just as what is nothing has none to number. So the void does not move[k] in time, so it does not move[k]. He added 'perhaps' because what is said is indubitable of the light and rare in the primary sense; but what they call the light is not light in its own nature, but by being mixed with void, so that if it were conjoined with the void through its body, they would be incommensurable, but if with void through its void, voids are not incommensurable. So the rarer would be that which exceeds in void but is exceeded in body in a certain proportion. So if motion[k] of void is abolished, there would be no cause of motion[k] of a sort in which moving[k] things are moved[k] by the void because of

its own motion[k], as the hypothesis of the unseparated void claimed. Motion[k] of the void would be abolished, both for that reason – that motion[k] of void is absurd – and for the others.

> 217a10-21 Since we deny the existence of void, [but the rest of the problem is truly stated, that there will be no motion[k] unless there is condensation and rarefaction, or the universe bulges, or there will always be an equal amount of water from air and air from water – for it is clear that more air is produced from water. So it is necessary, unless there is compression, that either the next thing will be pushed out and make the end one bulge out, or somewhere else a similar quantity of water will be transformed from air, in order that the total volume of the universe should remain constant, or nothing moves[k]. For this will always happen when there is displacement, unless that be circular; but travel is not always circular, but also in a straight line] – for these reasons they would say that there is a void.

They say that if there is no void there is no rarefaction and condensation; but without rarefaction and condensation there will either be no motion[k] or the universe will bulge. He well says that the first hypothetical is unacceptable, since there is no void but there is rarefaction together with condensation. But he rightly agrees that the remaining problems are justified. If there is really no rarefaction and condensation, either there will be no motion[k] or the universe will bulge, or, if not these, then, necessarily, if water were to come from air, somewhere else an amount of air equal, not to the water, but to the air from which the water came, will come from an equal amount of water. He makes it clear that he does not mean, as commentators have understood it,[142] air equal in volume to the water, by saying that 'it is clear that more air comes to be from the water'. I know that those who take the 'equal' in that way will understand this as disproving 'there will always be an equal amount of water from air'. But this is not how he means it, but as explaining how the equality is to be understood, not as of what comes to be to that from which it came; e.g. if water comes from air here, elsewhere an amount of air comes to be, equal to that from which the water came here, from an amount of water equal to that which came to be here. Aristotle made it clear that he so understood 'equality' when he draws his conclusion and says: 'So it is necessary, unless there is compression, that either the next thing will be pushed out and make the end one bulge out, or
687,1 somewhere else a similar quantity of water will be transformed from air, in order that the total volume of the universe should remain constant, or nothing moves[k].' For he would not have added 'some-

where else' if he meant the water that comes to be to be equal to the air from which it comes.

How is it clear that 'more air is produced from water'? Is it clear from the rupture of the vessels through the transforming of the liquid into a current of air,[143] as one sees in the case of jars of perfume, continually attaining a greater volume through transformation.

He takes up the problem again, confirming that there must be rarefaction and condensation, if impossibilities follow for those who abolish it, in order finally to prove that rarefaction and condensation come about, not from interspersed void but through another cause, and thus to abolish the void as unnecessary to preserve these, together with motion[k].

It is worth noticing that those who hypothesise an equal volume of things transformed and state an imaginary thesis do not provide for all change[k], but only for growth and coming to be.

He has said that it is necessary, if there is no void, either that nothing moves[k], or, if something did move[k], that continuously, as one thing after another was pushed, either the universe would bulge and stick out, or that transformations should be equal. Now he adds another way in which it is possible, there being no void, for there to be motion[k] without any of the above happening. This is circular reciprocal exchange of position. For in the case of circular motion there is no need for bulging or equal transformation or rarefaction and condensation, because things moving[k] in a circle always maintain the same place and have no need for expansion. But having said this he rejoins the objectors, saying that, even if circular motion[k] escapes the problems, needing none of the above, still not all change[k], even of place, is circular. There is change[k] of place in a straight line, so that the genuine problem remains in this case, that without compression there will be no motion[k] upwards or downwards, or else the universe bulges, or that transformations should necessarily be equal. It was to avoid these problems that they posited the void, as freeing them from all such difficulties. This persuasive presentation of the problem was not theirs, but he pointed it out in coming to the aid of the argument, as 'they would say', I think, sufficiently shows, indicating that they did not say it but would gladly say it if someone were to exhibit such an approach to them.

217a21-b11 But we say that [there is a single matter of contraries, of hot and cold, and the other natural oppositions, and that what is actual comes from what is potential, and that matter is not separable, though its essence is different, and may be the same for colour and hot and cold. Also there is the same matter in both a large and a small body. This is clear, for when air comes from water the same matter has become something else, not by acquisition, but what it was potentially it has become actually, and water comes from air likewise. Sometimes the big becomes small, sometimes the small big. Now, in the same way, even if much air comes to occupy a smaller volume and becomes bigger from smaller, the potentially existing matter becomes both. As the same matter becomes hot from cold and cold from hot, being potentially both, so it can become more hot from hot, nothing hot entering the matter which was not hot when it was less hot; and as the periphery and curve of a greater circle, if it comes to be of a lesser circle, whether or not the curvature remains the same, the curvature has not come to be in what was not curved but straight, since difference in degree is not through the quality being intromitted; nor is it possible to find a quantity of fire in which there is no heat or brightness. So the earlier heat is related to the later in the same way. So too the perceived size or smallness of the perceived volume is not extended by the addition of matter, but because the matter is potentially both. So it is the same thing that is rare and dense] and there is one matter of both.

35 He has said that we deny the void, but that the other problems were well raised by those who raised them, that if there were no compression, one of the above-mentioned absurdities would arise. For either
688,1 there would not be motion[k] or the universe would bulge, or in the case of things increasing in size there would have to be equal transformation. Now he solves the problem, allowing that there is compression of bodies, but not admitting that for that reason there is also a void.
5 Instead he provides another explanation of condensation and rarefaction based on the substrate matter of bodies. Having taught about this in Book 1, he reasonably announces that he will state his argument on this basis, i.e. what is already agreed.[144] It was said that matter is the substrate of opposites; thus there is the same matter of hot and cold, and similarly of dry and wet and other opposites, being
10 transformed from the one to the other, since it is always actually one of the opposites and potentially the other. In existence the matter is never separate from one of the opposites, but in its own definition and in concept it is something in addition to these. It remains numerically
15 identical, but receives each of the opposites. So, just as there is numerically the same substrate matter of qualitative opposites, such as hot and cold, white and black, or sweet and bitter, so with those in quantity, or concerned with quantity, or including quantity; e.g. there is numerically the same matter receptive of the great and the small in the case of bodies. For, if matter had some size or other of its own, it would not receive the opposite to that which it naturally had,

neither a greater nor a smaller, just as if it were naturally white it would not receive black. But as it lacks size and body, like other forms, it receives also the opposites in size while remaining numerically identical. This is clear in the case of transformation from water into air; for no other matter is added, when air comes from water. For if the matter that once was water was potentially air, when it is transformed from potentiality to actuality the same matter which formerly was water becomes air while remaining numerically identical, and the volume which once was less becomes greater, without matter being mixed in or void entering. For it would not have been potentially air if it became air by some mixture. In the same way also, if water comes from air, it is transformed into a lesser volume without the departure of any matter.

He made this clearer by taking the air itself which becomes smaller from bigger and bigger from smaller, without the addition or departure of anything; but the air itself becomes denser or thinner since its matter is potentially both and is transformed from potentiality to actuality, thus changing from a greater to a lesser volume and from a lesser to a greater. But since it is not transformed from being sizeless to having a size, or the reverse, but from greater to smaller or from smaller to greater, which is similar to the more and less in the case of qualities, he proves in this way that in the case of qualities also the same substrate is thus transformed into the more and the less as into opposites. For as the same matter becomes cold from hot and hot from cold, being potentially both, so it becomes hotter from hot. For the transformation does not come about from some parts of the substrate formerly not being hot and now becoming hot, the less hot being so through the admixture of parts that were not hot. For when the body was less hot, so the whole matter was likewise, and when it became hotter, the whole similarly increased. For if some parts of it were hot when it is less hot, others when it is more so, the numerically same matter no longer becomes hotter and less hot, as it receives the more and the less, but something else hot is added to what was there before and is mixed with it. So, as in these cases, in the case of the body that becomes rarer and denser, and greater and smaller, there is the same matter that receives both, and nothing exterior enters or departs, as those who introduce the void say. They do not understand the nature of matter, that it receives opposites while remaining the same.

Having proved in the case of qualities that the identical substrate receives the more and the less, he proves it also in the case of shape, since, even if this is a quality,[145] it has much kinship with size and quantity. For if the periphery of a greater circle be bent into the figure of a smaller circle, it is not by certain parts of the periphery that formerly were not curved but straight becoming now bent that it is reduced to a smaller circle, but by the less curved becoming more curved. And whether the periphery that was formerly less curved but later became more so be called the same, or whether each is called different from the other, this is very clear, that curvature did not succeed from the straight, but that the less curved became more curved. He seems to have added 'different' for superabundance of proof. For, if someone does not like the periphery that once was less curved but now is more so being the same, but says that each is different, even so, a curve has not replaced a straight line, but the less curved has become more curved. Generally, more and less are not so regarded because the form is intermitted in what is called less but not in the more. For the form is not added to, but the whole is intensified or relaxed, like the warmer, the sweeter and the whiter. So thus in the case of size and smallness of perceived volume, the matter does not extend by acquiring an extra magnitude, nor contract by losing it, as those say who invoke the void, but the still identical substrate, which was potentially greater, has become actually greater from smaller, and what was potentially smaller has become actually smaller from greater. For this is what it is for a substrate to be receptive naturally of opposites, which he made clear by 'the potentially existing', not meaning that it is matter potentially, but that the matter is potentially these. As flame is both hot and bright in every part, so too it is also rare in every part, and not by a mixture with void. Similarly the former warmth in relation to the later admits being more and less, not by the addition or subtraction of something. For no form is added to or subtracted from, but the same form is intensified or relaxed. He called the enmattered volume 'perceived' to distinguish it from the conceptual, i.e. mathematical.[146] For it is the perceived and enmattered volume that increases and decreases.

It is worth noting that magnitude is regarded as a certain form and the small and great as formal differences in magnitude. For these proceed into matter as accounts that make the form determinate.[147] For as animals and plants are made determinate by being white and black, hot and cold, and the like, so by their own size. This is why the greatest and least size is determined for each animal.

217b11-12 The dense is heavy, the rare is light.

He has shown that it is possible for there to be rarefaction and condensation, even if there is no void, since the suitability of the matter suffices for the increase and decrease of bodily size. He next demonstrates how the difference in motions[k] in a straight line follows from being light and dense. For if the rare is light, but the light moves[k] upwards by its own nature, it is clear what becomes rare does so not through a void but through its substrate matter, will move[k] upwards since it is light together with being rare. Similarly, if the dense is also heavy, and the heavy moves downwards, we shall have the causes of change[k] of place in a straight line without need of a void.

> **217b12-16** Again, as the periphery of a circle [when contracted to a lesser, does not acquire another bit of curve, but that which was there is contracted, and any part of a fire one takes will all be hot,] so the whole is a contraction and expansion of the same matter.

This passage seems to say again what was said before, how what becomes more and less does so not by some addition or subtraction, in the case either of the curved figure or of the warmth of a fire or of increase and decrease in size, but as the same matter is transformed into opposites. It adds nothing to what has already been said on this matter, when he said 'just as neither the periphery and curvature of the greater circle',[148] etc., but is a paraphrase of this. So it is omitted in some copies as introduced from a marginal comment[149] (which is the character of its language); instead the proof which says 'the dense is heavy, the rare is light' is followed by the words 'for there are two cases on each side, both the dense and the rare' etc.

> **217b16-20** For there are two cases on each side, both the dense and the rare. [For both the heavy and the hard seem to be dense, and both the light and the soft the opposite, rare. For weight and hardness do not coincide] in the case of lead and iron.

He has said that the dense is heavy, the rare light. Now he adds that not only weight follows on being dense, but also hardness. Again in its turn, lightness does not alone follow from being rare, but also being soft. 'There are two cases on each side', both the dense and the rare, on one the heavy and the hard, on the other the light and the soft. We shall see the need for the addition of the soft and the hard in a little while, when by this he proves that the cause of the three other kinds of change[k] as well as that of place is matter. He reminds us that these do not always go together, even if they do in most cases, by the example of iron and lead. For though lead is softer than iron it is however heavier than it in the same volume and more ductile. Perhaps he did not add this pointlessly, but to show that ductility is itself a form which does not arise from the weaving in of more void. If it did, it would inevitably be lighter. Now this is seen not to be the case, since, though lead is more ductile than iron and softer, it is still heavier than it.

217b20-7 From what has been said it is clear [that there is no separate void, either on its own or in the rare or potentially, unless someone wants generally to call the cause of travel the void. If so, the matter of the heavy and the light, as such, would be the void. For the dense and the rare, through their opposition, bring about travel, and, by being hard or soft, make things liable, or not liable, to be affected, and thus are the cause] of becoming different, rather than of travel.

Of those who affirm the void, some said that it existed actually and separated from bodies, others that it existed potentially, as coming to be and perishing for bodies. Of those who said that it was separate and actual some said that it existed on its own and independently, outside bodies, such as those who place the void outside the universe, others that it was separate and actual, but not independent, being interspersed in bodies and dividing their continuity, this being the cause of rarity. So he proved that it existed neither as independent
692,1 nor interspersed in bodies. So in no manner does it exist as actual and separate. 'But "potentially", says Alexander, 'may mean that against which he directed his final arguments, which, by being compounded and mixed with bodies, was said to be the cause of their rarity and lightness. This, I think, would be the interspersed void. But it has been proved that the void does not bring about rarity.' That is what Alexander says.[150] But it is worth considering whether the potential void was said to be the cause of rarefaction and not that which was actually interspersed. So perhaps he calls the potential void that which does not yet exist, but which, by the separation of bodies, occurs from time to time and from place to place; just as we ourselves say that channels come to be within which there is always something thinner than the surrounding body. It is clear that if the actual void has no entry into things the potential will not have it either. For this potential is everywhere that which is naturally brought to actuality, unless the substrate denies it existence.

He has proved that there is absolutely no void, and, since those who posit the void do so as the cause of motion[k], he combines with the argument the correction available to him and reminds us that if they call the void whatever is the cause of change[k] of place, but the cause of this is the matter receptive of rareness and density, or, more exactly, of lightness and heaviness, matter may be called the void. For matter may be suitably so called.[151] If it were a substance, this would be the void, since it has been shown to exist, but is none of the recognised forms, but by abstraction is conceived as void of them all.

Since the rare and the dense each has two types – the rare has the light and the soft, and the dense has the heavy and the hard, as is to be expected – the rare and the dense are the cause of motion[k] in a straight line *qua* light and heavy, but, *qua* soft and hard, not of this, but are rather susceptibilities or resistances to transformation and becoming different. For the soft is so susceptible but the hard is the reverse. He immediately proved in this way that matter, not void, is the cause of change[k] in the form of alteration, since alteration follows from hardness and softness, and these from rarity and density, and these through matter. We now have to understand alteration as not only transformation in quality, but also that in quantity and that in substance.[152] For acting and being acted on are seen in all these forms. And see, now it has become clear why the antithesis of soft and hard was brought in, for it was so that he could in this way prove that matter is the cause of the other three kinds of change[k], qualitative, quantitative and substantial.

Having said that the matter of the heavy and the light was the cause of change[k], he added 'as such' because matter underlies all these differences, but is not the cause of change[k] of place through all 693,1
of them, but only through the antithesis of light and heavy.

217b27-8 So much for the distinction of the way in which the void exists from the way in which it does not.

At the beginning of the discussions about the void he proposed to enquire whether it exists or not, the way in which it exists and what it is. Now, also, concluding the discussion he says 'the way in which it exists and the way in which it does not', because as a separate or potential void it does not, if one treats the void as an interval deprived of body, whether actual or potential. If, however, one treats the void as matter that is the cause of motion[k], he agrees that in that way the void exists.

That, then, is Aristotle's treatment of the void. But Strato of Lampsacus tries to prove the existence of a void that separates all body, so that it is not continuous, saying that 'neither light, nor warmth, nor any other bodily power would be able to make its way through water or air or any other body. For how would the rays of the sun make their way to the bottom of a container? For if the wet did not contain ducts, but the rays forced their way through to it, the contents of the the containers would come to overflow, and some of the rays would not be reflected in the upper part, while some made their way to the bottom.'[153] But I think it possible to solve these problems in accordance with the Peripatetic hypotheses, according to which warmth and other bodily powers, and light, not being bodies, do not need a void interval as a support of their being and a passage way, but subsist in bodies without increasing their volume.[154] But, even if someone were to say that light is a body, and an enmattered body, judging, from the sublunar light of the sun, its reflections and refractions by solids, that it is mixed with passible matter, still it is possible to resolve the problem by rarefaction and condensation. For nothing prevents rare bodies like water and air from being compressed and yielding space for some of the rays to move down through them. But such rays as fall on denser parts are reflected back. So, as I said, one can thus perhaps solve these problems, without being prevented from abolishing the really existent void. But perhaps the independent interval receptive of bodies and having different suitabilities to their differences has not yet been abolished by what has been said. For in rarefaction and condensation there must be some space and interval beside that of the bodies, which receives what is rarefied. For this does not happen in the neighbouring body, but in the interval from which the neighbouring body moved by being pushed or compressed.

But Plato also can be seen to deny the interval that is really void and deprived of all matter, and he states the final cause of this. 'For', he says, 'the cycle of the universe, since being round it comprehended all kinds and naturally wished to be in communion with itself, encloses all things and allows no empty space to be left.'[155] For, in general, the actual existence of such a void subverts both the continuity and the interaction of the universe, and will have no need to exist. For what is the point of a place superfluous to the bodies that would occupy it, if every space is for the sake of its contents? Also, if one contains body, another does not, how are these assigned, there being no difference within the void? However, Plato himself can be seen to know the interval that is other than the bodies that are in it, and having a different body in it at different times, and he calls it void, when he says: 'Things that consist of the biggest parts left the biggest void in their composition, the smallest the least. But the introduction of compression pushes the small things together into the voids of big ones.'[156] A little later he made it abundantly clear that transported bodies did not take over some space empty of body, but one that formerly contained another body by mutual replacement, where he says: 'Again, when the fire escapes thence, since it does not go out into a void, the adjacent air is pushed and itself pushes the moist volume that is still easily mobile[k] into the place of the fire, mixing it with itself.'[157] It is clear that the seat of the fire is some space where the fire had been. That this is an interval of the same size as every body that comes into it we may recollect from what was earlier said about place. For things moving[k] from one place to another, whether in a circle or in a straight line must, as was said before,[158] move[k] within some interval. So if it is not in their own nor in that of neighbouring bodies, they will move[k] in some other separate from bodies. This is the interval that gives the clear and obvious conception of place and void.[159]

Notes

Abbreviations

ANRW: *Aufstieg und Niedergang der Römischen Welt*
CAG: *Commentaria in Aristotelem Graeca*
SVF: *Stoicorum Veterum Fragmenta* (ed. H. von Arnim)
ABG: *Archiv für Begriffsgeschichte*
CQ: *Classical Quarterly*
MH: *Museum Helveticum*
REG: *Revue des Études Grecques*
RhM: *Rheinisches Museum für Philologie*
SO: *Symbolae Osloenses*

The translator's notes are indicated by (J.O.U.). All other notes are by Peter Lautner.

1. The most natural translation of *kenon* as an adjective is 'empty', but, as a noun *to kenon*, 'the void'. For consistency, 'void' is used as a translation throughout. (J.O.U.)

2. We have two theses here which do not necessarily converge: (i) the void is place deprived of body and (ii) place is void interval receptive of bodies. Simplicius assigned both (i) and (ii) to the circle of Democritus (*in Phys*. 397,3-4; 571,24-5.28-9) and the Epicureans (571,24-5.28-9), as well as to the Stoics and certain Platonists (571,27.29-31) who thought of place as always containing bodies and therefore did not admit actual void. D. Sedley ('Two conceptions of the vacuum', *Phronesis* 27, 1982, 175-94) argues that Simplicius' reference to Democritus on the void as unoccupied place is derived from Aristotle's attribution which is, in turn, of dubious value. The earlier Atomists considered void an occupant of place rather than a species of place itself (p. 185) in contrast to Epicurus who, constrained by Aristotle's treatment in *Physics* 4, conflated it with place. For a different interpretation, see B. Inwood 'The origin of Epicurus' concept of void', *Classical Philology* 76, 1981, 273-85. Simplicius criticises (ii) in a different way from Aristotle in his *Corollary on Place*, 619,3-622,37.

3. Although it seems that the whole discussion of the void is directed against the Atomists, the term 'vessel' (*angeion*) is surely not derived from them. Rather, it is Plato in the *Timaeus* who calls space or room (*khôra*) receptacle (*hupodokhê*, 49A6, 51A5), nurse (*tithênê*, 49A6, 52D5) and mother (*mêtêr*, 51A4-5).

4. The 'what it is' (*ti esti*) at 646,4 is omitted from the manuscripts, only the Aldine edition has it. The Peripatetic Aspasius' remark is hard to justify and elsewhere Simplicius disagrees with it by saying that *pôs esti* refers to the accidental attributes (*huparkhonta*) of the object (520,26-521,2). W.D. Ross and

Philoponus also deny his identification, the former supposes *pôs esti* to mean 'in what sense does it exist', and *ti esti* 'what is its nature' (*Aristotle's Physics*, Oxford 1936, 581), while the latter thinks *pôs esti* refers to the way that it exists: either infinite and outside of the ordered universe or interspersed in bodies (*in Phys.* 610,24-5, *CAG* XVIII). The reason why Aspasius took *ti esti* as equivalent to *pôs esti* might be that for him both phrases refer to the same thing, the substance. He accepts that *ti esti* alludes to substance (*ousia*, cf. *in EN* 11,24, *CAG* XIX,1) and *pôs ti* may ask to which of the categories the void belongs, since *pôs ekhon* was the designation of one of the Stoic categories. The context in Simplicius reveals that, for Aristotle, the supporters of void clearly consider it substance, in the Aristotelian sense of the term (cf. 648,15). For Aspasius' activity, see H. Gottschalk, 'Aristotelian philosophy in the Roman world from the time of Cicero to the end of the second century AD', *ANRW* II 36, 2, Berlin-New York 1987, 1074-1179, esp. pp. 1156-8, partly reprinted in R. Sorabji (ed.), *Aristotle Transformed*, London 1990, 55-83; and P. Moraux, *Der Aristotelismus bei den Griechen*, Bd. II, Berlin-New York 1984, 226-93, esp. pp. 238-9.

5. *Phys*. 4.1, 208a28-9.

6. *Phys*. 4.9, 217b27-8.

7. What Aristotle takes as a criterion is not the common conception (*koinê ennoia*) but rather the *endoxon*, a view held by (i) the majority of people or (ii) by distinguished philosophers (*EN* 1.4, 1095a28-30; 1.8, 1098b27-9). Common conception is a Stoic notion, providing the starting point for further investigation (Plutarch, *de Comm. Not*. 1075E, *de Stoic. Repug*. 1076C; Cicero, *Acad*. 1.42; Sextus, *AM* 9.132; Alexander, *Mixt*. 217,2-9.). Cf. F.H. Sandbach, 'ENNOIA and PROLEPSIS in the Stoic theory of knowledge', *CQ* 24, 1930, 44-51; R. B. Todd, 'The Stoic common notions. A re-examination and reinterpretation', *SO* 48, 1973, 47-75; M. Schofield, 'Preconception, argument and God', in M. Schofield, M. Burnyeat, J. Barnes (eds), *Doubt and Dogmatism: Studies in Hellenistic Epistemology*, Oxford 1980, 283-309. Simplicius employs *endoxon* for signifying the opinion of a particular school (651,25), or uses it in the sense of 'plausible' (50,28, 476,25-8, 1227,33), contrasted to 'demonstrative' (*apodeiktikos*, 610,19), therefore *endoxon* (at 651,8 translated as 'common opinion') and *doxa* (opinion) have a similar meaning.

8. According to E.R. Dodds, Proclus, *The Elements of Theology*, Oxford 1963, 344-5 there are three usages of the word '*epitêdeiotês*' (here translated 'suitable in nature'): (i) inherent capacity for acting or being acted upon in a specific way, (ii) inherent affinity of one substance to another (cf. Iamblichus, *De myst*. 5.7), (iii) inherent or induced capacity for the reception of a divine influence (Porphyry, *Ad Marc*. 19, Iamblichus, *De myst*. 5.23, Proclus, *in Tim*. 1, 139,23). *Epitêdeios* (or the noun *epitêdeiotês*) as a term was invented by Philo the Megarian to signify the inherent quality of a body (Alexander, *in An. Pr*. 184,6-10; Simplicius, *in Cat*. 195,32), which need not necessarily be realised. The way in which Simplicius uses it may lack this terminological rigour and here we can take the term as a substitute for 'capacity' (*dunamis*) and can at best rank it under one of the groups (i)-(ii) mentioned by Dodds. For an examination of the term, see R.B. Todd, '*Epitêdeiotês* in philosophical literature: towards an analysis', *Acta Classica* 15, 1972, 25-35. Elsewhere, however, Simplicius gets closer to the original meaning, cf. 665,9 (unnoticed by Todd). See n. 78.

9. This testimony about Anaxagoras is not registered in D-K, obviously because it is nearly the same as Aristotle's report in *Physics* 4.6, 213b24-7 (59 A68 D-K), cf. 59 A115 D-K.

10. Mere conception (*epinoia monê*) has nothing to do with common conception, cf. 506,5-10 where it is paralleled with *phantasia*. There are studies of this phrase in preparation by Frans de Haas (Ph.D. dissertation, Leiden) and Victor Caston (Ph.D. dissertation Austin, Texas and London).

11. This is the only direct evidence for the early Atomists' assumption of extracosmic void, recorded as 67 A20 D-K.

12. Disciple of Democritus, 70 A8 D-K. We have a rather unreliable testimony (Aëtius II 1,3 = 70A7 D-K) saying that Epicurus' teacher was a certain Metrodorus. But there were several Metrodoruses and one of them, who came from Lampsacus, was also associated with Epicurus.

13. For a detailed account, see 651,25-652,17.

14. *Phys*. 4.6, 213b22.

15. The reliability of this report becomes questionable when we read in Epicurus (*Ep. Hdt*. 41-2) that the universe (*to pan*) is unlimited both in the multitude of the bodies and in the magnitude of the void (cf. Lucretius 2, 958-64). See R. Sorabji, *Matter, Space and Motion*, London and Ithaca NY 1988, 125-42.

16. Simplicius' appreciation (cf. also 70,11), and refusal, of Porphyry's reading is shared by W.D. Ross, op. cit., 582. This text is cited as 'fonte 43' for Porphyry's commentary on the *Physics* in F. Romano, *Porfirio e la fisica aristotelica*, Catania 1985, and treated on p. 60. The merit of Porphyry's explanation is that he, unlike Simplicius, clearly distinguishes the earlier Atomists from the Pythagoreans by attributing extracosmic void solely to the latter.

17. Reading *legontos* instead of *legontes*, presumably a misprint. (J.O.U.)

18. *antiphatikê diairesis* ('disjunction is of contradictories') may be of Stoic provenance and hints at the division into an affirmative and a negative formula, here into 'void' and 'not void' (cf. 42,16, 451,26-7). Simplicius considers it vague at *in Cat*. 405,25-407,14. Aristotle contrasts only affirmative formulas (like health and sickness, black and white) in *Cat*. ch. 11.

19. Argument of the form *tollendo tollens*; if *p* then *q*, but not *q*, therefore not *p*. (J.O.U.)

20. Despite the fact that the two clauses have been put between quotation marks by Diels, the language may show that we are dealing not with fragments but rather with a paraphrase of a portion of 30 B7 D-K (cf. *in Phys*. 104,5-6). Our text is otherwise not registered in D-K, nor in Reale (*Melisso. Testimonianze e frammenti*, Firenze 1970), but it is in R. Vitali, *Melisso di Samo: Sul mondo o sull' essere*, Urbino 1973, Test. CXV. Diels fails to include the occurrences of 'Melissus' into the *index nominum*. Simplicius' further explanation is characterised by Neoplatonist vocabulary. It is Proclus who considered diversity or otherness (*heterotês*) the ground for division, and for being other (*in Parm*. 1184,2-28; *in Eucl*. 6,5; *Theol. Plat*. 4.27, 79,18-23, 80,8-9; 30, 89,21-3, 90.5-7.14 ff. (Saffrey-Westerink); *in Crat*. 22,14-15). At 432,34-433,1 Simplicius describes *heterotês* as the ground for being different, in which the different participates during the process of differentiation (*heteroiôsis*), cf. *in Cat*. 74,28, 245,7-8, 308,15-16 and *in Phys* 692,28 (commenting on 217b26). For the development of this notion in the Neoplatonists up to Proclus, see W. Beierwaltes, 'Andersheit. Grundriss einer neuplatonischen Begriffsgeschichte', *ABG* 16, 1972, 166-97.

21. When the word *ekei* (here) refers to the intelligible world, as opposed to the world of becoming (*entautha*, here), I translate it with capital letters. (J.O.U.)

22. A portion of his fragment registered as 28 B7 D-K.

23. This remark of Alexander, along with the following ones, may have been part of his commentary on the *Physics*.

24. For Alexander's notion, see his *Mixt*. 223,34-224,24. At 224,5-6, the school which considers *pneuma* (breath) a void (*ti kenon*) may be the Pythagoreans. But cf. the note of R.B. Todd in his *Alexander of Aphrodisias on Stoic Physics*, Leiden 1976, 218.

25. Although 'monad', 'dyad' and 'triad' may at first sight appear to denote ideal numbers (*eidêtikoi arithmoi*), at 650,11-12 we are told that the void has no place in the realm of the Forms, and this same point bears, I think, upon the realm of ideal numbers as well, since their difference must be due to otherness. Later on (652,13-15), Simplicius says it is the void that separates the Forms in the bodily cosmos and is found between the monad and the dyad as well. The members of a series (*ta ephexês*) between which there is no medium are also distinguished by the void (651,33-652,1). But mathematical numbers constitute a series (*in Cat*. 138,32-4), therefore we are plausibly dealing with mathematical numbers here. Or, perhaps, Simplicius takes over Syrianus' view that there are 'physical numbers' which organise the physical world (*in Metaph*. 189,13, 192,2-3, *CAG* VI,1). For this notion, cf. D.J. O'Meara, *Pythagoras Revived: Mathematics and Philosophy in Late Antiquity*, Oxford 1989, 134-5. Our text may support the first assumption.

26. The clause bracketed by Diels may be Simplicius' own explanation. The beautiful mentioned here is elsewhere called 'the archetypal beautiful' (*to arkhêtupon kalon*). There (637,4-6) a threefold division is sketched: (i) the Form '*kalon*' HERE (*entautha*) in the world of becoming, (ii) the enmattered form (*enulon kalon*) or the composite of matter and form which is called so, and (iii) the archetypal Beautiful. The first has a special arrangement (*suntaxis*) towards (*pros*) the other forms and this role corresponds to that of time and place (637,7-8). The archetypal Beautiful is therefore to be taken as a regulative principle and the Form of beautiful HERE is derivatively also endowed with this dignity. Simplicius states that beautiful is correlative to good (*agathon*, *in Cat*. 181,20-1) which is the highest 'entity' in Plato's *Republic* 509B9, cf. *in Cat*. 70,14 ff. The sentence then says that all Forms exist through the union THERE and on the ground of their relation to the beautiful.

27. Diels gives 258B sqq. The term here is not *heterotês*, but *thateron*. Elsewhere, Plato is speaking of a kind of not-being which is not opposite to being, but only different from it (257B3-4). At the end of this text, however, Simplicius tends to interpret void as not-being in the way Plotinus did. The latter also considered matter as not-being because it is completely indefinite and purely potential, cf. *Enn*. 2, 4.10.3 ff.; 5.5.5 ff. Aristotle mentions '*heterotês*' at *Phys*. 3.1, 201b2, which may be an allusion to the Old Academy, cf. *Phys*. 7.4, 249b23.

28. Throughout the discussion of the Pythagoreans, as, earlier in the case of Parmenides, it appears that the expression 'there is nothing there' is treated as equivalent to 'there is a void there' and also as having its normal English meaning 'there is nothing there'. 'Nothing', again, becomes 'that which is not', so the void interval in the physical world becomes identical with metaphysical 'not-being'. (J.O.U.)

29. Strato fr. 61 (Wehrli), probably a passage from his book on the void (cf. D.L. 5.59 = fr. 18). Simplicius is silent about whether Strato refuted the argument from compression. He may have connected compression with the existence of void interspersed in bodies, cf. F. Wehrli (ed.), *Die Schule des Aristoteles V, Straton von Lampsakos*, Basel 1950, 58; H.B. Gottschalk, 'Strato of Lampsacus: some texts', *Proceedings of the Leeds Philosophical and Literary Society*; Literary and Historical Section 11.6, Leeds 1965, 131. At 693,14 Simplicius reports that he accepted the argument from the percolation of light (fr. 65a, Wehrli), which entails that he agreed with the existence of void in the bodies themselves. But this was the core of the argument from compression too. See D. Furley, 'Strato's theory of the void', J. Wiesner (ed.), *Aristoteles. Werk und Wirkung*, Bd. I, Berlin-New York 1985, 594-609, repr. in his *Cosmic Problems*, Cambridge 1989, 149-61.

30. *Phys*. 4.6, 213a23 ff., Simplicius ad loc. 647,20-648,11.

31. For Aristotle, lightness is a tendency to move upwards, as weight or heaviness is a tendency to move downwards. These are equally perceptible to touch. (J.O.U.)

32. Reading *autous* instead of *autois*. (J.O.U.)

33. 'Problem' in the technical sense of a question of a form 'Is A B or is it not B'. See Aristotle, *Top*. 1.3, 101b29. (J.O.U.)

34. *An. Post*. 2.1, 89b24 ff. (J.O.U.)

35. Aristotle, *Phys*. 4.3, 210a14 ff., Simplicius *ad loc*. 551,1 ff.

36. It is mostly the Presocratics who were called 'naturalists' (*phusiologoi*) and first perhaps by Aristotle (*Metaph*. 1.5, 986b14, 8, 990a3; *DA* 3.2, 426a20; *Phys*. 3.4, 203b15, 5, 205a27, 6, 206b23; 4.7, 213b1) who once ironically says that Plato's *Timaeus* also 'naturalises' (*DA* 1.3, 406b26). One of Chrysippus' works treating the Presocratics was entitled 'The ancient naturalists' (cf. D.L. 8.187). The epithet implies originally not simply materialism but rather a distinction from the poets (cf. the case of Empedocles in Aristotle's *Poetics* 1447b18-19), a turn to the explanation of physical phenomena. Here Simplicius scornfully applies it to the circle of Anaxagoras, which is unjust although he may have shared the disappointment of Socrates (*Phaedo* 97B8-98B6) with Anaxagoras' explanations.

37. In Simplicius, the first syllogism is *Barbara*: Every heavy or light thing is tangible; Every tangible thing is body – Every heavy or light thing is body. The second is *Camestres*: Every heavy or light thing is body; No void is body – No void is heavy or light.

38. If everything that is is body and every body is tangible having weight or lightness, then, since they lack such attributes, mathematical objects must be void or else they do not exist at all, which is absurd.

39. The MSS support 'not filled'.

40. At *DA* 48,21 ff., Alexander mentions again that the air seems to be void, but rejects the idea in *Mant*. 125,19-25. Alexander's arguments fit in well with the debate about whether qualities were of a bodily nature, as some Stoics claimed (*SVF* II, 380, 383, 388, 394, 410, 449, 467) or not (this was the view of Ps.-Galen's *De qualitatibus incorporeis* 462,2 ff. K, re-edited by M. Giusta, 'L'opusculo pseudogalenico *Hoti hai poiotetes asomatoi*', *Memorie dell' Accademia delle Scienze di Torino. Classe di Scienze Morali, Storiche e Filologiche*, Ser. 4a, n. 34, Torino 1976). Alexander himself attacks the Stoic view in the relevant section of the *Mantissa* (122,7-125,4), using the example of the air in a wineskin (124,13-20, cf. 125,21).

41. We know this reading from Alexander alone.

42. The Greek reads *treis ekhôn skhedon diastaseis*, which I would translate as 'having roughly three dimensions', were that not absurd. (J.O.U.)

43. *Phys*. 1.3, 269b29, Simplicius, *in Cael*. 62,5-63,24.

44. Aristotle may be thinking of Plato's *Timaeus*. Simplicius explicitly attributes this view to the Platonists after Aristotle.

45. For Aristotle, heavy and light are linked to the four sublunar elements and determine their vertical movement, while hot, cold, fluid and dry constitute the defining qualities of these elements, e.g. the air is fluid and hot (*GA* 2.2-3). In contrast to that, heavenly bodies made up of a peculiar matter, the *quinta essentia*, are free of such attributes and their only change is the everlasting rotation (*Cael.* 1.2, 269a2-b10) that is the perfect movement. The heavenly body (singular) is the whole mass of transparent spheres which carry the heavenly bodies.

46. That is, they consider void to be prime matter, i.e. that which cannot be a particular thing, nor a bodily substance. On such an interpretation void would be the ultimate subject of properties and in need of determination, just as prime matter is.

47. *Phys.* 4.1, 209a4-9, 2, 209b26-8, 4, 211b29-34.

48. A syllogism in *Cesare*: No matter is separable, all void is separable, therefore no void is matter. (J.O.U.)

The emphasis is on the 'separable' (*khôriston*). But if void were separable only in thought or according to definition, not in actuality, then the counter-arguments are rather flimsy. Prime matter, like void, is only separable in thought from things of which it is the matter. Aristotle, and Simplicius, may exploit the fact that '*khôriston*' is used in two senses, first as 'separable' and then as 'separated'. The latter would clearly imply actual void and actual prime matter.

49. See *Corollary on Place* 618,21-4. If Simplicius thought of the Old Academy he might have had in mind Crantor's commentary on the *Timaeus*. This work was known to Iamblichus and Proclus as well. For further references to these philosophers, see 571,30, 601,23.

50. According to these arguments, change of place would be impossible since local motion requires some interval which is, however, supposed to be place. Simplicius discusses Aristotle's objection at 659,5 ff. As regards place, cf. *Phys.* 4.4, 211b5-212a2, Simplicius ad loc. 570,16-578,13.

51. *Cat.* 5a15 ff. (J.O.U.)

The body having only quantitative marks – that is, magnitude – was considered by them mathematical body, extension without qualities (*apoion diastaton*). This view was taken by Herminus (ap. Simplicium, *in Cat.* 124,33-6), a Peripatetic of the second century AD and the teacher of Alexander (cf. H.B. Gottschalk, op. cit., 1987, 1158-9; P. Moraux, op. cit., 361-99, esp. pp. 370-1), and by Lucius (ap. Simplicium, *in Cat.* 125,13-16), perhaps a Platonist who might have published his work between AD 160 and 180 (cf. W. Capelle, 'Lukios', *RE* XIII, 1791-7). If this is so, it might have been accessible to Alexander.

52. The above passage is awkward to translate. 'The interval of a body' means 'such an interval as a body could occupy', while 'bodily interval' means 'interval consisting in a body'. (J.O.U.)

53. Reading *ton topon* instead of *ho topos*. (J.O.U.)

54. 650,1-4.

55. 650,4-12.

56. *Phys.* 1.3, 186a18. (J.O.U.)

57. *Phys.* 1.3, 186a17

58. Fr. 63 (Wehrli). With the aid of this experiment Strato refutes the argument from local motion to the void and his refutation is not necessarily bound to the supposed lack of compression. We are just told that exchange of place (*antimetastasis* or *antiperistasis*) need not involve void, pace H.B. Gottschalk, op. cit., 1965, 131. This argument is of course directed against the existence only of void on its own, not of that in a body (659,10).

59. In *GC* 1.5, 321a9-29 Aristotle distinguishes the transformation of elements from ordinary alteration, see W.D. Ross, op. cit., ad loc. (J.O.U.)

60. The first *anapodeikton* of the Stoics (*modus ponens*): if *p* then *q*, but *p*, therefore *q*. The wording also is reminiscent of them.

61. *DA*. 2.4, 415b26. (J.O.U.)

62. Aristotle claims at 321b16-28 that the form is what grows and is nourished for it is not all parts of the matter that grow, but the form (321b32-5, cf. *DA* 2.4, 416a9-18). Form is a measure (321b24-5) that remains the same but measures successive parts of the water. In this passage, Alexander uses the phrase *ti pote esti* ('what it is in fact', 661,2.4) that in Aristotle, as well as in Alexander, alludes to *ho pote on* referring to substrate (*hupokeimenon*) free of any relational aspects. See A. Torstrik, '*ho pote on*. Ein Beitrag zur Kenntnis des aristotelischen Sprachgebrauchs', *RhM* 12, 1857, 161-73.

63. It appears here that the lemma immediately following – the passage from the *Physics* of Aristotle – is treated as an integral part of the text. I have noticed no other place in which this occurs, and some scholars believe the lemmata to be editorial additions. The word 'saying' [*legôn*], unless deleted, seems to undermine this view. (J.O.U.)

64. This passage, and the discussion by Simplicius, are easier to understand if one remembers that the problem is that of the assimilation of food so that it actually becomes part of the body, rather than being merely interspersed through the body. Aristotle says this is a difficulty for everyone, which is not solved by supposing that growth occurs through food being stored up in void areas of the body. (J.O.U.)

65. Themistius uses the proverb in the same context at *in Phys.* 127,6-7 (*CAG* V,2), cf. also Philoponus, *Corollary on Place* 577,3 (*CAG* XVII), Aeschylus, fr. 139 Nauck(2).

66. The *reductio ad absurdum* consists of two main parts: either the food is not body, and then the assumption of void is useless as food does not penetrate body, or it is body, but then it is absurd to suppose void to exist for bodies are growing in every part, and if so, either body interpenetrates body, which he thinks impossible, or the whole body will be void, that is body and void will be the same.

67. *GA* 1.5, 321b26 ff.

68. Fr. 81 (Wehrli), *Die Schule des Aristoteles VIII, Eudemos von Rhodos*, Basel 1955.

69. Fr. 62 (Wehrli), cf. 652,19-25 (fr. 61) where, at 652,22 ff., Strato is reported as explaining attraction by attributing a power to the magnet in virtue of which it draws the material out of the pores of the iron, which in turn will be recovered from another iron, and so on. For the structure of the testimony, see H.B. Gottschalk, op. cit., 1965, 132, n. 1.

70. Simplicius' demonstration relies on Peripatetic principles. Additional premiss or additional assumption (*proslêpsis*) may take its origin in Aristotle (*An. Pr.* 2.8, 59b19), and propositions formed from an indeterminate middle and two determinate extreme terms are called by Theophrastus 'by *proslêpsis*' (testimonies 110A-D, Fortenbaugh et al., see also Simplicius *in Phys.* 107,22 which explains 107,12-16 = Theophrastus, testimony 155C). 'Potentially' may hint at this fact too and not only at the fact that Aristotle actually left his thesis unproved. Later authors, like Sextus (*PH* 2.149) use the term to signify the minor premiss in the Stoic form of argumentation, cf. *SVF* III 269, D.L. 7.82. Simplicius seems to retain the Peripatetic usage, cf. *in Cat.* 143,17-20; *in Phys.* 107,12-16, 171,21-2, 490,1-4,

515,32-5, 664,8-10, 698,17, 932,21-3, 947,14-21, 1014,3-9, 1027,3-10, 1262,27-8, 1300,2-8.

71. A *modus ponens* again. The wording ('the first, so the second') also recalls the Stoics.

72. Aristotle's argument is surely not formally valid. It appears to be: If void is the ground of motion then void exists; but it is not the ground of motion, so it does not exist. This, however, is the fallacy of denying the antecedent. Perhaps that is why Simplicius calls the hypothesis of the void pointless rather than false. (J.O.U.)

73. *aition*, translated as 'ground', refers to the efficient cause and *sunaition* to a cooperating cause. Elsewhere, at 26,5-7 and 316,24-8, we are told that *aition* comprises the paradigm and the productive and final causes, while '*sunaition*' signifies matter, shape (*eidos*) and instrument. The source of this distinction is the transformation of the Aristotelian notion of cause by the Stoics into a substitute for the agent. This view was taken over by some Neoplatonists as well, like Porphyry (ap. Simplicium, *in Phys.* 11,6-7), Iamblichus (ap. Simplicium, *in Cat.* 327,6-17), Proclus (*El. theol.* 75; *in Tim.* 1, 1,24-4,5, 261,14-18; *in Parm.* 1059,3-19), Simplicius (see also *in Phys.* 326,25-33, *in Cael.* 467,21-5) and Philoponus (*in Phys.* 5,11-16, 186,18-21, 241,3-5). For the transformation in the Stoics, see M. Frede, 'The original notion of cause', in M. Schofield, M. Burnyeat, J. Barnes (eds), op. cit., 217-50, reprinted in his *Essays in Ancient Philosophy*, Oxford 1987, 125-51, who claims that *sunaitia* bring about effects in cooperation with other causes in contrast to the *aition* in the proper sense which produces effects by itself. Simplicius seems to insist on the concept of *sunaitia* as auxiliaries. Without the intervention of an efficient cause matter and shape are incapable of creating anything, however much they may collaborate with each other. On the other hand, if they cooperate with an *aition* they presumably play only a subservient role in this process.

74. Many of Aristotle's objections at 214b17-215a24, all based on the *ou mallon* (why this rather than that?) argument, were answered by the Stoics, Chrysippus (ap. Plutarch, *de Stoic. Rep.* 1054B-1055C) and in his wake Cleomedes (10.15-12.5), by supposing centripetal inclinations (*neuein*) of the elements and a holding power (*hexis*) which holds things together. Both are due to the *pneuma* that is supposed to have a centripetal motion by which it keeps together the cosmos, cf. R. Sorabji, op. cit., ch. 9, esp. pp. 148-53. For the argument 'why this rather than that', see S. Makin, 'Buridan's ass', *Ratio* 28, 1986, 132-48. He distinguishes two variants of the argument: (A) is of the form 'there is no more reason to say that p than to say that q', and (B) is 'there is no more reason for p (to be true) than for q (to be true)'. The arguments reported by Simplicius in this section (664,7-29) belong to (B).

75. *Phys.* 3.5, 205a10. (J.O.U.)

76. Reading *menei* (rest) for *merei* at 664,20. See *homoiôs de oude menei* in the next line. (J.O.U.)

77. This is Aristotle's view, cf. *Phys.* 4.4, 212a6. (J.O.U.)

78. The suitability (*epitêdeiotes*) of fire and water is to move in a definite direction, and in using the *ou mallon* argument Simplicius does not deny the existence of such capacities; rather he points to the void as an impediment to these capacities being realised. Thus the original meaning of the term seems to be saved (cf. n. 8). See also 690,20.

79. Reading *soma ti* with Ross, instead of *somati* as in Diels' text. This is probably a misprint: Diels has *soma ti* at 665,34. (J.O.U.)

80. *Phys.* 4.4, 211b20 ff. (J.O.U.)

81. At the beginning of *Corollary on Place* Simplicius offers a short doxography on how place was conceived, cf. 601,1-24.

82. 664,15-29.

83. Diels points out that the words quoted are from *Phaedo* 109A4-5, but that the point is made also at *Timaeus* 40B and 62D12-63B1. (J.O.U.)

The reference to 40B is less convincing. This thought is exploited in Democritus (68 A44 D-K) and probably in Anaximander (12 A1, 11, 26 D-K) to establish the middle and equipoised position of the Earth. For the way the argument works, see S. Makin, op. cit.

84. Again, Alexander is our only witness for this omission in some copies.

85. 'Normal state' (*proêgoumenê huparxis*) denotes here the elementary and most fundamental way in which something exists or can exist. For *huparxis* as existence, see *in Cat*. 218,23-4, 365,24, 421,20-1; *in Phys*. 490,4, 542,27, 1117,20. As existence or state to which a proper *energeia* belongs, *in Cat*. 301,25-6, 333,2-6. To understand the meaning of '*huparxis*' here it may be helpful to take into account the passage at *in Phys*. 194,1-10 where Simplicius says that it is not substance (*ousia*) that at 189a11 Aristotle contrasts with the other categories, but the whole *huparxis* of the physical 'subsistence' (*hupostasis*) in change. This *huparxis* is physical, generated and enmattered. As regards 667,16, this meaning of the term may be emphasised through '*proêgoumenê*' which means 'primary' and refers to a thing's existence on its own (*kath' hauto*, see *in Cat*. 128,11-12, 141,2-9, 428,30; Syrianus, *in Metaph*. 164,13-16,33, 165,7-21; Porphyry, *in Cat*. 103,13, Simplicius, *in Phys*. 4,27, 249,23, the last three places are cited also by A. Grilli, 'Contributo alla storia di *proêgoumenos*', *Studi linguistici in onore di V. Pisani*, Vol.1, Brescia 1969, 409-501, who confirms this meaning. But see, Syrianus, *in Metaph*. 110,28-9 where '*ta proêgoumena huparkhonta*' means primary attributes). For a similar construction (*proêgoumenê apodeixis*), see 668,12. The whole problem is of Stoic origin for which, see P. Hadot, 'Zur Vorgeschichte des Begriffes Existenz: *hyparchein* bei den Stoikern', *ABG* 13, 1969, 115-27; against this interpretation V. Goldschmidt, '*Hyparchein* et *hyphistanai* dans la philosophie Stoicienne', *REG* 85, 1972, 331-44, and see also A. Graeser, 'A propos *hyparchein* bei den Stoikern', *ABG* 15, 1971, 299-305.

86. In D.L. 7.76-81, the second *anapodeikton* (*modus tollens*) has the form: 'If it is day, it is light. But not: it is light. Therefore not: it is day.'

87. This *modus ponens* repeats the one at 667,27-30.

88. This may still be Alexander's own addition. Aristotle argues in this way neither here nor at *Physics* 5.6, 230a18-231a17.

89. *Phys*. 4.8, 214b28-215a14.

90. '*proêgoumenos*' shows that Simplicius refutes forced motion in itself, regardless of any link to its counterpart, natural motion.

91. *Timaeus* 59A1-4, Burnet gives a text different from Simplicius' at 59A2 by putting *eti* after *onta*. For that version, cf. 694,19-21 and n. 157.

92. The former explanation of projectile motion, that the air displaced in front of the projectile goes round and pushes it from behind, is here replaced by the second explanation which Aristotle finally endorses in *Physics* 8.10, 266b27-267a20. According to this, the air behind the projectile is made into an 'unmoved mover'. Philoponus ridicules this solution, cf. his *in Phys*. 641,16-26 (*CAG* XVII). See R. Sorabji, op. cit., 144-5, 228-9. E. Hussey objects to Sorabji's interpretation in 'Aristotle's mathematical physics: a reconstruction', in L. Judson (ed.), *Aristotle's Physics*, Oxford 1991, 213-43, esp. p. 231, n. 40 (see also the whole section on projectile motion, pp. 230-6 and 243) and insists on the possibility of an 'impulsion' theory in Aristotle which makes room, long before Philoponus, for an impetus to be implanted in the missile, not merely in the air. Simplicius also seems to waver as to why it is the air and not the missile on which motion is impressed, cf. *in Phys.* 1349,26-9.

93. Aristotle does not answer this question in the *De Motu Animalium*.

94. Philoponus was one of them, cf. *in Phys.* 644,16-22 (*CAG* XVII).

95. *Physics* 5.6, 230b10-231a4, Simplicius ad loc. esp. 913,10-916,31. See also 671,29-30, 677,20 ff., as well as *in Cael.* 212,9-30 (*CAG* VII).

96. The notion of 'holding power' (*hexis*) is characterised in Cleomedes 10.15-12.5 and Chrysippus ap. Plutarch, *de Stoic. Rep.* 1054E-1055A, ap. Alexandrum, *Mixt*. 217,32-218,10, cf. n. 74. Alexander's objection is not against the possibility of empty intervals between bodies, but against extracosmic void. Cleomedes does not raise the question where the ordered universe is lodged perhaps because the question itself implies an infinite regress to him, just as it does for Aristotle. For what would be the place of the universe, if place is defined as 'limit of the container' (212b6) and if there is nothing outside the universe which could contain it? Aristotle's answer to this puzzle became controversial in late Antiquity. For this problem in the Stoics, see Chrysippus' view ap. Plutarch, *Stoic. Rep.* 1054F (with H. Cherniss' footnote *a* on p. 585 of his edition, Cambridge/Mass. 1976) and Plutarch's query at 1055A ff., which is similar to Alexander's. See also the Introduction.

97. It is not clear whether Aristotle is talking of one body in different situations at different times or of two bodies simultaneously in different conditions. Simplicius seems to interpret the text in the latter way, so I have done the same. (J.O.U.)

98. 678,10-680,9.

99. For Aristotle a point is where a line may be divided. But the two resulting segments comprise the whole of the line. It does not lose a portion, which is the point of the division. (J.O.U.)

100. Philoponus attacks the argument by establishing that the absence of any resistance in a void eliminates not the need for time, as Aristotle supposes, but only the need for *extra* time, for the body to complete its motion. If, in the absence of resistance, motion took no time at all, Aristotle would have to admit that the outermost heaven required no time to revolve (*in Phys.* 690,34-691,5, *CAG* XVII), see R. Sorabji, op. cit., 145-8, 283-4.

101. This follows from the definition of time itself as well, according to which time is a number of movement, or belongs to movement in another way, cf. *Physics* 4.11, 219b1-2, 220a24-6, 221a13-4, b2, 11, 22-6. The reason for Simplicius to make this apparently pointless remark may be that by his time the link between physical movement and time had already weakened. Iamblichus spoke of static time (ap. Simplicium, *in Cat.* 354,18-25, 355,3-21.26-32; *in Phys.* [*Corollary on Time*] 787,12-23, 793,2) and Proclus followed him (*in Tim.* 3, 32,7-16;), cf. R. Sorabji, *Time, Creation and the Continuum*, London 1983, 37-42.

102. Adding a comma after *pheretai* and deleting comma after *kenou*. (J.O.U.)

103. Alexander's reason may be that in the preceding argument (215a24 ff.) Aristotle was talking about the difference in a medium constituted by thinner or denser things and he turns to considering void explicitly as medium only now. The problem of how this transitional passage is to be placed inheres in the text of the *Physics*. Philoponus sides with Alexander (*in Phys.* 656,18 ff.).

104. The asterisked passage is a translation of a conjecture by Diels to fill an apparent lacuna. (J.O.U.)

105. *Phys*. 4.8, 216a5. (J.O.U.)

106. *Phys*. 4.8, 215b13, Simplicius ad loc., esp. 673,5-17.

107. The lemma ends with the words *hôsper khronos pros khronon* which are at line 216a7, an inadvertence. The next lemma begins at 216a12. (J.O.U.)

108. Throughout, *A* is an object in motion, *B* is a medium traversed, as are *D* and *F*, which is usually, but not always a void. *C* is a time, as are *E* and *G*. *H* is a portion of *D*. (J.O.U.)

109. Simplicius had access to the works of some commentators (*exêgoumenos* or *exêgêtês*) living before the time of Alexander, who wrote on aspects of Aristotle's natural philosophy, even if not a full-scale commentary on the *Physics* as we have it now. These are: Theophrastus, Eudemus of Rhodes (his own *Physics*), Strato of Lampsacus, Andronicus of Rhodes, Nicolaus Damascenus, Boethus of Sidon, Adrastus, Aspasius and Galen. It cannot be excluded that Alexander too was acquainted with all these texts, along with other Peripatetic treatises, for natural philosophy was favoured in the school, H.B. Gottschalk, op. cit., 1987, 1154 ff.

110. See the passage quoted from Alexander at 675,5-10.

111. The text cited by Simplicius is different from his quotation of the same text (216a15) at 675,14-15. Here we read '*touto adunaton*', at 675,14-15 '*sumbêsetai touto to adunaton*'. Bekker and Ross follow the latter version.

112. For *antikinêsis* ('contrary motion'), see 670,27-8 and n. 45. Aristotle ascribes counter-movement to every mover except the first (*GA* 4.3, 768b19; cf. *Phys*. 8.5, 257b23-4, *Cael*. 1.5, 272b4, *Mem*. 2, 453a26-7). Simplicius also asserts that every body must be exposed to this change (*in Phys*. 419,19-35; cf. 1017,10, 1239,24-32; *in Cael*. 186,23, 212,9-30, 366,7-8, 395,27-396,5, 396,28, 405,35, 418,25-6 where counter-movement is said to be the principle of generation, *arkhê tês geneseôs*). If *antikinêsis* belongs only to bodies, it cannot be attributed to the void.

113. Does this passage imply that for Simplicius space is void so far as its own nature is concerned, though it is always in fact filled by bodies, as was the view of his Platonist predecessors and Philoponus? The evidence is not quite clear but he seems to talk of actual void as an interval to be traversed by bodies, and so defends motion through void on the ground that the difference in speed may come from the moving bodies themselves because their impulses are not equal. If different speed is due to different impulses there is no need to admit that in the void all bodies must have the same speed. But the solution gets stuck half-way. The difference of speed does entail that motion through void as such will not be timeless, i.e. instantaneous, but if so, void ought to be in some ratio to the full since Simplicius maintains the Aristotelian principle of the relation of speed to resistance. In contrast, Philoponus rejects the whole edifice by pointing out that the revolution of the heavens encounters no resistance, but that its speed nevertheless remains finite, cf. Philoponus' *Corollary on the Void*, 681,17-30, 690,34-691,5. Furthermore, Philoponus relies on experiment when he says that if you drop a body doubled in weight from the same height, its fall will not be twice as fast, cf. 683,5-25. Both objections are based on experience, *pace* E. Hussey, op. cit., p. 242 n. 59. For Philoponus, see R. Sorabji, op. cit., 1988, 145-7.

114. 620,16-27, as an adherent of this view, Syrianus is also cited at 628,26-34.

115. The relation of speed of fall to a body's density is denied by Philoponus, *Corollary on the Void* 683,16-25.

116. There is no evidence in the MSS for this reading at 216a17.

117. The impulse (*rhopê*) Alexander is speaking of is simply the weight and not impetus in any technical sense.

118. Reading *autois* for *autais* at 679,16. (J.O.U.)

One could defend *autais* by taking it as referring to *phora* and so translate: '... in the case of travel, size ... '.

119. As he does at *in Metaph*. 36,22-4 (*CAG* I), Alexander here uses the Atomist term '*allêlotupein*' ('hit into') for describing the mutual collision of the atoms, cf. Aëtius I 12.6. The main problem posed by him is how objects of familiar size come into being if, moving downwards at the same speed, atoms do not hit or get entangled with each other. However, as Simplicius adds, this holds true only if atoms are in eternal motion for their temporary rest might allow entanglement. To justify this thesis we ought to assume that atoms are always moving parallel. But Alexander ignores the swerve which is responsible for their inclination, a notion missing in the letters of Epicurus (though see *Ep. Herod*. 62), but found in Lucretius who claims that were there no swerve (*declinare*, 2.221-4, see also Cicero, *De Fato* 22) atoms would not collide. Furthermore, atoms are also capable of moving upwards by 'squeezing out' (*ekthlipsis*, ap. Simplicium, *in Cael*. 267,29-268,4, Lucretius 2.191-205). Both theses were posited in order that there might be no obstacle to objects of familiar size, or *kosmoi* to come into being. The reason why objects large enough to be visible move at different speeds is that their constituent atoms may zig-zag.

120. With the statement that impulse is the only ground (*aitia*) for the natural travel of bodies Simplicius gets close to the position of Philoponus, who allows only internal causes for natural motion (*in Phys*. 384,18-385,4, 678,22-9, 679,27-31). This insight of Simplicius remained isolated and was not exploited any more in the commentary, and what is more, see 287,29-288,6. On the natural motion in Philoponus, see M. Wolff, 'Philoponus and the rise of preclassical dynamics', in R. Sorabji (ed.), *Philoponus and the Rejection of Aristotelian Science*, London 1987, 84-121; id., *Fallgesetz und Massebegriff*, Berlin/New York 1971, 38-87, and *Geschichte der Impetustheorie*, Frankfurt am Main 1978, 147-52.

121. Simplicius' objection is again similar to that of Philoponus provided that the addition to the difference in natural impulse is not infinitely great, as may happen in the case of the void. Although not excluding this case clearly, he might suggest that speed in a void does not increase infinitely but will vary with impulse.

122. Beginning of a *modus ponens*.

123. *Phys*. 4.4, 211b19-21, 4.6, 213a31.

124. *Phys*. 4.9, 216b22. (J.O.U.)

125. Reading *eiper mê eiê kenon*. Diels has *eiper eiê kenon*. (J.O.U.)

126. *salpingôn*. This word is always translated as 'trumpet'; but it is very strange to speak of a trumpet as being a reed-instrument. (J.O.U.)

127. *Phys*. 4.1, 209a4 ff.

128. This fits well Aristotle's interpretation of Plato's view in the *Timaeus* (*Phys*. 4.2, 209b17, see also the criticism at 4.4, 211a19-25). Simplicius also seems to equate the quantitative character of a body with the place this body occupies, and disregards the various senses in which one thing can be said to be 'in' another.

129. As a result of the explanation at 808,23-9, *pathos* can in a certain sense be taken for attribute (*sumbebêkos*), cf. the paraphrase from Alexander's commentary at 757,28-9.

130. 216a23-4. (J.O.U.)

131. 649,4 ff., esp. 650,18-29, 652,19-21.

132. This is all that we know of Xouthos the Pythagorean (33 D-K). He may indeed have thought that the universe bulges or swells (*kumainei*), that is, motion may occur without rarefaction and condensation. See also Themistius *in Phys.* 135,20-136,2. Philoponus only knows that he was a sophist (*in Phys*. 671,6).

133. Reading *sumbainei*. Diels has *sumbainein*, presumably a misprint. (J.O.U.) One could defend the infinitive by supposing that it depends on *phasin*, the main predicate of the whole passage.

134. The section began with the promise that only the void scattered in bodies would be treated (682,24), but first the impossibility of extracosmic void has to be proved again.

135. *in Phys*. 135,26-7 (*CAG* V,2). Simplicius' detailed analysis is found at 687,35-689,15.

136. The translation reproduces an anakolouthon in the Greek text. (J.O.U.)

137. The expression *to huph' hou* is meant to indicate the productive cause both in Aristotle (*phusis*) and Plato (*poioun*), see the classification by Porhyry ap. Simplicium *in Phys*. 10,25-11,15, cf. *in Cael*. 299,23-4. For the antecedents, see H. Dörrie, 'Präpositionen und Metaphysik', *MH* 26, 1969, 217-28, repr. in his *Platonica Minora*, München 1976, 124-36.

138. I have translated *hupokeimena*, usually translated as substrate, as 'burdens'. I suspect that the text should read *huperkeimena*. (J.O.U.)

139. Arguing for Simplicius' residence at Harran (Carrhae), where this commentary is also said to have been written, M. Tardieu refers to a similar sentence at *in Cael*. 525,10-13 in which Simplicius claims to have seen these inflated skin rafts on the nearby river Habur, see his *Les paysages reliques. Routes et haltes syriennes d'Isidore à Simplicius*, Louvain-Paris 1990, 71 ff., 134-47, 157.

140. This is similar to the thesis Simplicius accepts at 649,31-2, that the ground for the natural motion of bodies is internal.

141. He may be thinking of the Atomists, though they only assert that it is the atoms that are falling downwards. To infer from atomic fall to the motion of whole compound bodies would require the demonstration of the parallel motion of the atoms, which would therefore be responsible for the downward tendency of the bodies as well. But Atomist texts contradict this assumption, cf. n. 119.

142. e.g. Themistius, *in Phys*. 135,26-7, but not Philoponus who attacks Themistius at his *in Phys*. 671,10-672,17.

143. Omitting *hoti* with F at 687,4 and *ek* before *tês* with *a* at 687,5. (J.O.U.)

144. *Phys*. 1.9, 142a20 ff.

145. Repeated at 82,6-7. Aristotle introduces shape (*skhêma*) as a fourth genus of quality at *Cat*. 8, 10a11, contrary to *DA* 3.1, 425a18 where he says that it is a sort of magnitude (*megethos ti*). This ambivalence leads Simplicius to consider shape close to quantity. Although once we are told that shape is different from, and posterior to, quality (*in Cat*. 219,12-22), on the whole he remains faithful to the classification in the *Categories*. Shape admits no degree (*in Cat*. 112,11-12, 285,11-12.18-19, 285,34-286,1, 289,4) and its varieties are, e.g., triangularity, hookedness (*grupotês*) and the circle (*in Cat*. 226,9-11, 289,10), cf. also *in Cat*. 21,13-20, 48,18-19, 207,3, 220,35-221,1, 226,27-227,7, 261,20 ff. (reference to Iamblichus at 262,12-19) and 284,9-10. As all the examples are geometrical figures there is some reason to associate shape with quantitative features.

146. *noêtos*, usually meaning 'intelligible', is translated here as 'conceptual' because it refers to the realm of mathematical entities. Proclus also speaks of two kinds of volume: mathematical and physical (*in Tim.* 2, 25,3-4) or bodily (*in Tim.* 1, 300,14-15) and connects them to extension (*in Tim.* 1, 178,21-2). There remains the question whether this realm depends on our thinking or not. Simplicius offers no explanation of it.

147. *sumplêrôtikoi logoi* are defining characteristics that complete the essence of the thing in question, cf. 268,1-6. What is *sumplêrôtikos* is in this context essential, not additional or adventitious (*epeisaktos*, *in Cat.* 49,4-7) nor transcendent (*exêrêmenos*, *in Cat.* 80,1-2).

148. *Phys.* 4.9, 217b2-3.

149. The passage is missing in Themistius, and see W.D. Ross, op. cit., p. 594 ad loc.

150. Philoponus agrees with Alexander in looking on 'potentially' as standing for 'intermixed' (*enkekramenon*, *in Phys.* 701,22).

151. Just the prime matter, of course, which is deprived of all qualifications. Aristotle discussed it in *Metaph.* 7.3, 1029a1-27.

152. Simplicius claims that Plato too called quantitative change alteration (*alloiôsis*), cf. *in Cat.* 428,31-429,2, 429,13 ff.

153. Fr. 65a (Wehrli). The argument was unforeseen by Aristotle whose detailed examination of compression is found in *Physics* 4.9. D. Furley (op. cit., p. 155) claims that the only innovation made by Strato concerning void was the theory of microvoid and, e.g. by pointing out that microvoid-theory is compatible with *horror vacui*, ascribes it to Strato against the recent objections by M. Gatzemeier, *Die Naturphilosophie des Straton von Lampsakos*, Meisenheim am Glan 1979, 94-7. His views are accepted by L. Repici, *La natura e l'anima. Saggi su Stratone di Lampsaco*, Torino 1988, 123.

154. In *DA* 2.7, 418b7-14 Aristotle concludes that light is the actuality of the transparent (*energeia tou diaphanes*). In other words, light is what changes a medium that is potentially seeable-through into being actually seeable-through. Alexander also insists on this view in his *De Anima* 43,8-11; *Mant.* 124,9-10, 144,29. He devoted a separate treatise to prove that light is not a body, cf. *Mant.* 138,4-139,29.

155. *Timaeus* 58A4-7. (J.O.U.)

156. *Timaeus* 58B2-5. (J.O.U.)

But Plato postulates rectilinear corpuscles of shapes which ought to leave gaps between each other. The 'voids of big ones' means the voids between the bigger corpuscles.

157. *Timaeus* 59A1-4. The text is different at 668,34-669,2, cf. n. 91.

158. 620,16-27, 678,2, see also the quotation from Syrianus at 628,26-34.

159. At this point Simplicius moves on to a discussion of Aristotle's treatment of time, which immediately follows that of the void. (J.O.U.)

English-Greek Index

be able: *dunasthai*
abolish: *anairein*
abolishing: *anairesis*
absolutely: *haplôs*
abstraction: *aphairesis*
absurd: *atopos*
absurdity: *to atopon, atopia*
accept: *dekhesthai, homologein*
account: *logos, apodosis*
accretion: *proskrisis, proskrinomenon*
actuality: *energeia*
actually: *energeiāi*
add: *epagein, prostithenai*
addition: *diadosis, prosthêkê*
additional premiss: *proslêpsis*
admit: *sunkhôrein*
admixture: *mixis*
agree: *sunkhôrein*
air: *aêr*
allow: *ean*
alter: *alloiousthai*
alteration: *alloiôsis*
always: *aei*
animal: *zôion*
antecedent: *hêgoumenon*
antithesis: *antithesis*
argue: *sullogizesthai*
argument: *logos, epikheirêma, epikheirêsis, kataskeuê*
ash: *tephra*
atom: *atomos, atomon*
attack: *antilegein*
attract: *helkein*
attraction: *helxis, holkê*
attribute: *huparkhon, sumbebêkos*

ball: *sphaira*
be: *einai, huparkhein*
become: *gignesthai*
before, be: *proüparkhein*
beginning: *arkhê*
belief: *pistis*
bent: *kurtos*
bodiless: *asômatos*
bodily: *sômatikos*
body: *sôma*
book: *biblion*
boundless: *apeiros*
breath: *pneuma*
bulge: *kumainein*
bulk: *onkos*
burn: *kaiein*

call: *eipein, phanai*
can: *dunasthai*
carry: *okhein*
cause: *aitia, aition*
  cooperating cause: *sunaition*
  efficient cause: *poiêtikos, to huph' hou*
  final cause: *telikê aitia*
certainly: *pantôs*
change(n): *kinêsis, metabolê*
change(v): *ameibein, kineisthai, metaballein, metapiptein*
change of place: *phora*
channel: *poros*
circle: *kuklos, periodos*
clear: *dêlos, phaneros, saphes*
clearly: *dêlon, saphôs*
cold: *psukhros*
colour: *khrôma*
come about: *sumbainein*
come about, come to be: *gignesthai*
coming to be: *genesis*
commentator: *exêgêtês*
common: *koinos*
compel: *anankazein*
compose: *suntithenai*
composition: *sustasis*
comprehend: *sumperilambanein*
compress: *pilein*
compression: *pilêsis*
conceive: *ennoein*

concept: *ennoia, epinoia*
conception: *ennoia, epinoia*
common conception: *koinê ennoia*
conceptual: *noêtos*
conclude: *sumperainein, sunagein*
conclusion: *sumperasma, sunagetai*
condensation: *puknôsis*
condition: *pathos*
confirm: *kataskeuazein*
congregate: *sunizanein*
congregation: *sunizêsis*
conjunction: *sundesmos*
consequence, consequent: *sunêmmenon, hepomenon*
contain: *ekhein, parekhein, periekhein*
container: *angeion, angos*
continuity: *sunekheia*
continuous: *sunekhes*
contract: *sustellein*
contraction: *sunagôgê*
of contradiction: *antiphatikê*
contribute: *sumballein*
conversely: *anapalin*
by conversion: *antistrepsas*
corresponding: *analogon*
cosmos: *kosmos*
criterion: *kriterion*
cube: *kubos*

define: *horizein*
definition: *horismos, logos*
demonstrate: *apodeiknunai*
demonstration: *apodeixis*
prior demonstration: *proapodedeigmenon*
dense: *pakhumeres, pakhus, puknos*
density: *puknotês*
deny: *ou phanai*
departure: *to apienai, to existanai*
deprived: *esterêmenon*
destroy: *dialuein*
determinate: *horismenos, peperasmenos*
determine: *diakrinein, horizein*
determining account: *logos sumplêrotikos*
differ: *diapherein*
difference: *diaphora*
different: *diaphoros*
dimension: *diastasis*
dimensional: *diestôs*
diminution: *meiôsis*
discover: *heuriskein*
disjunctive: *diairetikos*
dissolve: *dialuein*
distinction: *diakrisis*
distinguish: *diakrinein, diorizein*
diversity: *heterotês*
divide: *diairein*
divisible: *diairetos*
division: *diairesis*
double: *diplos*
downwards: *katô, epi to katô*
draw conclusion: *sunagein*
dyad: *duas*

earth: *gê*
easy: *rhaidion*
easily: *rhaidiôs*
empty: *kenos*
end: *peras, telos*
enmattered: *enulos*
enquire: *zêtein*
equal: *isos*
examine: *episkeptesthai*
example: *paradeigma*
exceed: *huperairein, huperballein*
excess: *huperokhê*
exchange: *metastasis*
exchange place: *antimethistanai*
exchange of place: *antimetastasis*
exist: *huparkhein*
existence: *hupostasis, to einai*
explain: *aitiologein*
explanation: *aitiologia, aition*

feature: *pathos*
feed: *trephein*
fill: *plêroun*
filled: *plêres*
find: *heuriskein*
fire: *pur*
first: *prôtos*
follow: *akolouthein, hepesthai, parakolouthein*
food: *trophê*
force: *bia*
forced: *biaios*
form: *eidos*
full: *mestos, peplêromenon, plêres*

generality: *hoi polloi*
generally: *holôs*
genus: *genos*
gradually: *kat' oligon*
grow: *auxanesthai*
growth: *auxêsis*

happen: *sumbainein*
have: *ekhein*
heavenly: *ouranios*
heavens: *ouranos*
heaviness: *barutês*
heavy: *barus*
HERE: *entautha*
hypothesis: *hupothesis*
hypothesise, form hypothesis: *hupotithenai*
hypothetical argument: *hupothetikos*

idea: *ennoia*
illustration: *paradeigma*
impede: *empodizein*
imperceptible: *anaisthetos*
impetus: *rhopê, rhumê*
have impetus: *rheptein*
impossible: *adunatos, atopos*
impossibility: *to adunaton*
inanimate: *apsukhos*
inclination: *rhopê*
incline: *klinein*
incommensurable: *asumblêtos*
incorporeal: *asômatos*
increase: *epidosis*
indemonstrable: *anapodeikton*
indeterminate: *aoristos*
infinite: *apeiros*
inseparable: *akhôristos*
intelligible: *noêtos*
intentional: *proairetikos*
intersperse: *diaspeirein, paraspeirein*
be interspersed: *enspeireisthai*
interpret: *akouein*
interpretation: *exêgesis, hermêneia*
interval: *diastêma*
introduce: *eisagein*
inverse: *antistrophos*
irrationally: *para logon*
by, in itself: *kath' hauto, en hautôi*

kind: *genos*
kinship: *koinônia*
know: *idein*

later: *husteros*
length: *mêkos*
light: *kouphos, leptomeres*
lightness: *kouphotês*
likeness: *homoiotês*
limit: *peras*
limited: *peperasmenos*
line: *grammê*
local motion: *phora*

magnitude: *megethos*
make: *poiein*
mathematical: *mathêmatikos*
matter: *hulê*
mean: *akouein, hermêneuein, sêmainein*
meaning: *sêmainomenon, sêmasia*
medium: *to metaxu, to di' hou*
middle: *mesos*
mix: *mignunai*
mixture: *mixis*
mobile: *eukinêtos*
moist: *hugros*
monad: *monas*
moon: *selênê*
motion: *kinêsis, phora*
contrary motion: *antikinêsis*
move: *kineisthai, pheresthai*
movement: *kinêsis*
must: *anankaios*

name: *onomazein*
natural: *pephukos, phusikos, kata phusin*
nature: *phusis*
be of a nature: *pephukos*
contrary to nature: *para phusin*
necessary: *anankaios*
necessity: *anankê*
need: *deesthai*
not being, what is not: *mê on*
notion: *ennoia*
nourish: *trephein*
nourishment: *trophê*
number: *arithmos*
numerically: *kat' arithmon*

obstruct: *empodizein*
occupy: *katekhein*
opinion: *doxa*
opposite: *enantios, enantiôsis*
otherness: *heterotês*
  becoming other: *heteroiôsis*
own: *oikeios*

part: *meros, morion*
participate: *metekhein*
particular: *idios*
particular thing: *tode ti*
pass through: *khôrein*
passing away: *phthora*
pebble: *psêphis*
perceptible: *aisthêtos*
perception: *aisthêsis*
periphery: *periphereia*
place(n): *topos, khôra*
place(v): *tithenai*
plant: *phuton*
point: *stigmê*
portion: *meros, morion*
posit: *tithenai*
position: *thesis, hupothesis*
possible: *dunatos, endekhetai*
postulate: *thesis*
potentially: *dunamei*
power: *dunamis*
holding power: *hexis*
preserve: *phulattein*
primarily: *proêgoumenôs*
principle: *arkhê*
prior, be: *proêgeisthai*
privation: *sterêsis*
problem: *aporia, aporoumenon, problêma, zêtoumenon*
projectile: *rhiptoumenon*
proof: *apodeixis, deixis*
proper: *oikeios*
proportion: *analogia*
prove: *deiknunai*
push: *ôthein, epôthein, exôthein, proôthein, sunôthein*

qua: *hei, katho, kathoson*
quality: *poion, poiotês*
quantity: *poson, posotês*

rare: *manos*
rarefaction: *manôsis*
rarity: *manotês*
ratio: *logos*
  in inverse ratio: *antestrammenôs*
rational: *logikos*
reasonably: *eikotôs*
receive: *dekhesthai*
receptacle: *khôrêma*
receptive: *dektikos, khôretikos, hupodektikos*
refute: *anairein, elenkhein*
relation: *skhesis*
relational: *ta kata skhesin*
remain: *menein, histasthai*
replace: *epeiserkhesthai*
  mutually replace: *antiperiistasthai*
  mutual replacement: *antiperistasis*
resistance: *antitupia*
resistant: *antitupos*
retain: *phulattein*

say: *eipein, legein, phanai*
scatter: *diaspeirein*
sea: *thalassa*
see: *horan, idein*
seem: *dokein, eoikenai, phainesthai*
self-defeating: *empodizein heauton*
self-evident: *autopistos*
sense: *aisthêsis*
sensible: *aisthêtos*
separable: *khôristos*
separate: *apokrinein, dialambanein, khôrizein*
series: *ta ephexês*
shape: *skhêma*
share: *metekhein*
show: *deiknunai, endeiknusthai*
simply: *haplôs*
size: *megethos*
skin: *askos*
slow: *bradus*
solution: *lusis*
solve: *dialuein, luein*
soul: *psukhê*
space: *khôra*
spatial: *topikos*
speak: *eipein, legein*
special nature: *idiotês*
of the same species: *homoeides*
speed: *takhos*

state: *legein, phanai*
stone: *lithinos, lithos*
subsist: *huparkhein*
substance: *ousia, hupostasis*
substrate: *hupokeimenon*
subtraction: *aphairesis*
suitable: *epitêdeios, prosphues*
suitability: *epitêdeiotês*
sun: *hêlios*
superabundance: *periousia*
surround: *periekhein*
surround: *to perix*
susceptible to change: *eupathes*
sympathy: *sumpatheia*
syphon: *harpax*

take: *lambanein*
tangible: *haptos*
THERE: *ekei*
throw: *rhiptein*
thin: *leptos*
thing: *pragma*
think: *hêgeisthai, noein, oieisthai*
thinness: *leptotês*
thought: *noêma*
thesis: *thesis*
time: *khronos*
timeless: *akhronos*
together: *hama*
touch(n): *haptesthai*
touch(v): *haphê*
transform: *metaballein*
transformation: *metabolê*
transmission: *diadosis*
travel: *diienai, diodeuein, pheresthai*
true: *alêthes*
truth: *alêtheia*

unchanging: *akinêtos*
underlie: *hupokeisthai*
understand: *akouein*
undifferentiated: *adiaphoron*
undifferentiatedness: *adiaphoria*
union: *henôsis*
universal: *katholou*
universe: *holotes, kosmos, to pan*
unseparated: *akhôristos*
upwards: *anô, epi to anô, pros to anô*
moving upwards: *anophoron*
use: *khrasthai*

vegetative: *phutikê*
vessel: *angeion*
void: *kenôma, kenos, kenotês*
voids of: *diakena*
volume: *onkos*

war: *thermos*
warmth: *thermotês*
water: *hudôr*
way: *tropos*
weight: *barutês*
well: *kalos*
white: *leukos*
  turn white: *leukainesthai*
whole: *holos*
  the whole: *to holon, holotês, to pan*
wholly: *pantelôs*
  which wholly is: *to pantelôs on*
wine: *oinos*
wish: *boulesthai*
witness: *marturion*
word: *onoma*

Greek-English Index

adiaphoros, undifferentiated, 664,24.28; 665,8; 666,21.30; 667,27; 670,10.20; 671,8; *adiaphoria*, undifferentiatedness, 664,24
adiarthrôtos, vague, 653,7
adioristos, open-endedness, 658,30; 659,6; vaguely, 653,33; 654,10; indefinitely, 661,9; indeterminately, 661,18; open-endedly, 658,33; *eipein*, vagueness of the statement, 654,4-5; *lêphthênai*, vague meaning, 656,1
adranes, impotent, 669,21
adunatos, impossible, 645,24; 649,20; 650,26; 651,4; 658,7.8; 669,21.24.25; 671,20; 672,12; 673,13.22; 674,2.8.22; 675,13; 676,34; 677,8; 678,9.28.30-1.34; 679,25; 680,11.13; 681,16; 682,7; 684,3.21.25; 685,28; cannot be, 649,16; 650,31; 670,26; 679,12; not possible, 669,19; *to adunaton*, impossibility, 673,4.13; 674,26; 676,35; 677,11; 680,11.24.31; 687,8.36; absurdity, 675,15
aei, always, 657,25.27; 659,11.22; 667,25; 677,24; 686,28; 687,21; 688,10; 692,10
aêr, air, 647,19 *passim*; 648,29; 652,31; 653,30.33; 654,36; 655,2 *passim*; 656,11.18-19.21; 660,6.12.21-5; 661,19; 668,27.28.31; 669,1.4.6.9; 670,33.34; 671,29.35; 672,3.6; 673,2; 674,12.14; 676,21; 678,5; 679,8; 680,34; 681,7.9; 683,37; 684,1.3.5; 686,24 *passim*; 687,1.3.4; 688,23-33; 693,13.27; 694,20
agein, bring, 675,2; 692,13; reduce, 680,32
agnoeisthai, be unaware of, 658,34; not understand, 689,14-15
ainigma, riddle, 652,7
aisthêsis, sense, 654,23.35; 655,6.30.31; perception, 681,13
aisthêton, perceptible, 647,33; 654,12.15.26.36; 655,6 *passim*; 657,8; perceived, 689,34; 690,9-10
aitiasthai, explain, 650,28; 685,23; account for, 685,25; criticise, 659,4; invoke, 689,35
aitiologein, explain, 669,14-15; give explanation, 679,12; provide explanation, 688,5; *aitiologia*, explanation, 678,27
aition, cause, 646,10-11; 652,17; 663,7.10; 670,33; 671,14.25; 676,34; 677,19; 678,10.14.24; 679,15.29-35.38; 680,4.7.27; 684,27.29.31; 685,10-17.25; 686,12; 687,9; 690,26; 691,15.32; 692,4.16.18.26.30.39; 693,1.9; explanation, 679,4.31; reason for, 653,24; 657,34; agent, 651,31; ground, 658,12.17-20.30; 659,7; 661,9; 663,16.19-29.33-4; 664,3.6.26; 665,12; 666,16; 670,32; *telikê a.*, final cause, 694,3; *aitian eipein*, say why, 649,26-7
akhôriston, inseparable, 648,18.26; 658,24; 681,27; 684,14; unseparated, 657,21.27; 686,14
akhronos, timeless, 671,21; 672,12; 673,5.27; 677,13.34
akinêtos, unchanging, 650,6; 658,33; 659,1.17.19-20; do not move, 666,24
akolouthein, follow, 662,22; 674,3.4; 676,6; 679,22-3; 684,8-9; 690,22;

692,31; *akolouthos*, consequent, 670,17
akouein, hear, 647,23; 679,37; take to mean, 650,21; understand, 651,29-0; 654,18; 661,18; 684,1-2; 686,26.28; 692,32; interpret, 658,6
akribes, accurate, 665,18; precise, 683,5; adv., exactly, 692,19; *to a.*, exactness, 672,31; *akribôs*, accurately, 656,25
aktis, ray, 693,15.18.27
alêtheia, truth, 683,2; *alêthes*, truly, 646,33; true, 660,23; *alêthôs*, truly, 667,31; 682,8
allêlotupein, hit into, 679,18
alloiôsis, alteration, 658,31.34; 659,5; 660,12.24; 683,10.19; 692,30-2
alloiousthai, be altered, 658,32; be alteration, 660,20; 683,21
ameibein, change, 678,17; *topous*, take over each other's places, 659,9
anairein, refute, 647,21; 663,6; 664,32; abolish, 652,29; 670,8.21; 677,21; 686,12.14; 687,8.10; 693,33; attack, 655,8; eliminate, 683,6,7,33,34; disprove, 666,35; 668,10.23; *anairesis*, destruction, 683,6; abolishing, 693,30
anaisthêton, imperceptible, 655,23-4.28-9
anaklasthai, be reflected, 693,18; *anaklasis*, reflection, 693,24
anakrouein, collide, 671,29-30
analogia, proportion, 672,38; 674,6; 676,20.24; equality, 684,1; *kat' analogian*, proportionally, 685,34; *analogon*, corresponding, 647,30; be in proportion, 675,16
analuein, give answer, 663,4
anamphisbêtêton, indubitable, 686,6
anankaios, indispensable, 645,22; necessarily, 678,33; 680,34; probative, 676,18; necessary, 660,17; 662,5; 680,12; 687,10; force, 658,25; validity, 663,13; adv., with force, 658,26
anankazein, compel, 662,19; require, 663,4-5
anankê, is necessary, 649,12; 660,2; 664,5; 686,34; 687,14; is probative, 676,10; must, 651,3; 667,32; 677,28; 683,31; 684,7; 694,26; will, 670,12; necessarily, 668,17; 676,19; 684,2; 687,27; necessity, 658,21; 667,13; 668,17; 676,19; 682,3; inevitably, 670,16
anapalin, conversely, 674,18; the reverse, 677,20
anapherein, refer to, 646,8-9
anapheresthai, travel up, 685,27
anapnein, breathe in, 651,27
anapodeikton, indemonstrable, 668,1
anatrepein, overrun, 655,3; overthrow, 661,22
anereunasthai, examine, 646,34
angeion, vessel, 646,1.23; 651,10.19; 659,13.23-5; 662,15.20; 687,4; container, 681,13; 693,15.17
angos, pail, 651,15; container, 681,12
anienai (from *aniêmi*), relax, 689,32; 690,9
anisos, unequal, 649,32-4; 677,5
anisotakhes, of different speed, 671,24-5; difference of speed, 672,1; 679,16; unequal speed, 677,30; 678,10.14; *anisotakhôs*, at unequal speed, 679,15
anô, above, 664,12; 665,9; 667,21-2.28; upwards, 669,32; 684,27.31; 685,4 *passim*; 690,23.25; upper, 693,18; *epi to*, upwards, 663,31-2; 671,27; 678,21; 685,20.25; 687,26; *pros to*, upwards, 664,29
anôphoron, moving upwards, 671,32; travel upwards, 685,9; leading upwards, 684,29
antanaklasthai, be reflected back, 693,28-9
anteipein, attack, 647,30-1; reply, 680,20
anteirêken, replied, 663,3; spoke against, 683,1
antestrammenôs, in inverse ratio, 676,4
anthrôpoi, people, 647,32-3
pros antidiastolên eipein, distinguish, 690,9-10

antigraphos, copy, 665,26; 667,4; 691,3
antikeimenos, opposing, 646,36; opposite, 685,20
antikineisthai, be in opposite motion, 671,29; be in contrary motion, 677,21
antikinêsis, contrary motion, 677,20
antilegein, attack, 656,30
antilêptos, perceptible, 654,23
antimetastasis, exchange of place, 659,21; 683,29
antimethistasthai, exchange place, 659,7-8.15; take over, 659,25
antiperiistasthai, be mutual replacement, 668,26.29; be reciprocal exchange, 687,18; mutually replace, 669,16; *antiperistasis*, mutual replacement, 668,33; 694,18; reciprocal exchange, 683,30
antiphatikê, of contradictories, 649,15
antipheromenon, be in opposing travel, 678,13
antiphraxis, refraction, 693,24-5
antistrepsas, by conversion, 667,31
to antistrophon, the inverse, 674,1
antithesis, antithesis, 692,36; 693,2
antitupoun, resist, 647,19; *antitupia*, resistance, 656,8; *antitupos*, resistant, 647,22; 656,11.19.21-2
apagein, drive, 656,26
apagogê, reduction, 667,20
apantan, use (the doors), 648,34; come about, 665,33-4
aparithmêsis, 653,21
apatasthai, be mistaken, 647,31; 648,5.6
apeinai, be in absence, 668,24
apeiros, unlimited, 645,26.27; boundless, 663,16; 664,10.16.18.22-3.32; 667,18-19.22; 671,1.5-6; infinite, 664,15; 682,2; be no bound, 667,26; *ep' apeiron*, without limit, 645,27-8; *ad infinitum*, 682,18; without end, 669,26; endlessly, 668,19; *eis apeiron*, endlessly, 670,12.16; be infinite regress, 670,26
aphairein, take away, 647,29
aphairesis, subtraction, 690,31; abstraction, 692,22
aphesis, sending off, 669,11
aphian, let out, 647,29
aphôrismenôs, exclusively, 664,12; 669,32-3
aphorizesthai, be distinguished, 647,3
(***to***) ***apienai***, departure, 688,32; substraction, 690,7
apistia, disbelief, 646,16.20.22
apodeiknunai, demonstrate, 663,9; prove, 664,6; show, 666,16
apodeixis, proof, 668,12.15.21-2; 680,16; 689,26; demonstration, 671,18; 682,3
apodekhesthai, accept, 655,12; justify, 686,22
apodidonai, speak of, 656,23-4; provide, 679,5; 680,20; 685,21; give, 685,11
apodosis, account, 655,34; 656,14.16.18; 657,10
apokremannusthai, hang from, 652,25
apokrinein, separate, 682,27; 691,26-8.30; 692,2; 693,7
apoleipein, leave, 648,21
apollusthai, perish, 691,27
apoluein, break, 666,3; free from, 687,28
apoporeuein, give off, 662,30
aporein, raise the problem, 662,7; 687,35-6; *ho aporon*, objector, 687,22; *êporêmenon*, problem, 686,22; 687,25
aporia, aporia, 654,13; puzzle, 661,14-15; 662,6; problem, 661,25-6; 662,2-3.18.23-5; 663,19; 687,7.23.29; 688,3; 693,26
(***to***) ***aporon***, absurdity, 654,21.29
aporoumenon, problem, 655,4
aporrein, run out, 662,9-10; flow out, 662,13
apoteleisthai, be achieved, 650,28
apsukhos, inanimate, 660,26
araios, thin, 660,5

arguros, silver, 671,35; 678,21
arithmos, number, 653,3.34; 672,14-16.20.22.26; 673,8.10; 674,9.22; 675,25; 686,4; 688,15.18.23.26; *kat' arithmon*, numerically, 688,13, 689,9
arkesthai, suffice, 653,29; be able, 653,33; be sufficient, 661,19; be based, 667,13
arkhê, cause, 669,9; principle, 679,14; beginning, 673,32; 683,2; 693,5; *to ex arkhês*, the former, 649,22; the original, 649,28.30; *tên arkhên*, at the start, 679,27-30
arkhêsthai, begin, 670,2; 673,31
asapheia, unclarity, 662,6
asaphês, unclarity, 665,21
askos, wineskin, 647,23; 655,4; skin, 650,20.22-4; 651,12.14.16; 659,32; 685,1.3
asômatos, bodiless, 653,26; 673,2; incorporeal, 662,19; not be body, 693,21
asumblêton, incommensurable, 684,10; 686,1.9-10
athroos, concentrated, 668,29; 669,6.13; condensed, 669,11
atmis, vapour, 662,30-1
atomon, atom, 679,14-15.23
atopon, absurd, 649,31.34; 662,17; 672,22; 675,13.20.25; 677,5; 681,20.25; 684,6; 686,15; (*to*) absurdity, 654,4.10.31; 655,38; 656,26.33; 664,22; 666,6.9; 672,2; 675,3.12; 677,6.18.25; 678,18; 679,22; 680,24; 684,9.26; 685,5.9.24.27; impossible claim, 673,27; impossibility, 676,31; problem, 687,27; *atopia*, absurdity, 670,29
augê, ray, 693,17
aulos, shawm, 681,7
autopistos, self-evident, 649,12
auxêsis, growth, 648,27-8; 649,9; 650,30.33; 651,2-3; 660,13-14.23.25-6; 661,4. *passim*; 662,10; 683,9; 687,13
auxesthai, grow, 651,7; 660,11.18.19.27.28.30; 661,2-3.5.31-4; 683,12-13.27; increase, 679,28
axiologos, noticeable, 684,20-1
axios, worthy, 660,26
axioun, require, 662,4; think it worth, 666,33

ballein, insert, 662,3
baros, weight, 653,31; 655,2; 656,12.15; 671,33; 672,3.10.32; 678,20; 679,16.26.33.36
barus, heavy, 653,32.35; 655,35.37; 671,33.35; 678,16.26.29.33; 679,2; 681,23.30; 685,10.17.22.23; 690,18.26; 691,5.9-10.12.22-3; 692,25-6; 693,2; (*to*) *baru*, heavy thing, 650,28; *to baruterôi einai*, to be heavier, 678,29
barutês, weight, 679,27.34; 681,31; heaviness, 685,19.20-1; 692,19
basanizein, test, 646,36
biâi, by force, 667,6; 668,28; 669,4.24.25; forcibly, 669,22; forced, 669,26; 693,16
biaios, forced, 667,9.12.14.32.34; 668,10.13.21-2; 669,13; *biaiôs*, forcibly, 669,31; *meta bias*, forcible, 669,11; with force, 669,12
biazesthai, force, 668,14.17-18.24; 681,10
biblion, book, 659,5.17; 688,6
boêthein, come to the aid, 662,33
bôlos, clod, 678,16
borboros, mud, 671,28
boulesthai, wish, 648,29; 672,39; 675,32; 676,22; 694,5; intend, 657,33; 666,17; claim, 650,6; be supposed, 654,1; regard, 566,18
braduteros, slower, 671,31; 679,2; more slowly, 677,28; 678,25
brakhus, small, 660,7; 661,20; 678,16.17; brief, 674,32

daktulos, finger, 660,9-10
dapanan, spend, 662,32
dedeigmenon, demonstration, 666,9-10
deesthai, need, 659,13; 669,9; 681,27; 682,16-17; 687,21.23; 690,27
deiknunai, prove, 646,7; 647,21;

649,10.17.22; 650,7; 651,1.5; 652,31; 655,37; 657,2.16.31; 660,13; 661,4; 662,1.23; 666,17.19; 667,8.18; 668,20.25; 669,24; 670,7.15; 671,13.18; 672,13.16; 673,21; 674,2.23 676,33; 678,18; 680,15.27.29.31; 684,22; 685,29.31; 687,9; 688,37; 689,16-17; 691,15.32; 692,6.29.36; 693,11-12; show, 647,17.22; 648,7; 655,18; 656,20; 657,15.19; 658,18.24-5.27; 659,6.11.31; 661,11.17; 662,4; 663,30; 664,3; 673,33; 682,8; 684,21.32.34; 690,19; 692,15.21; demonstrate, 675,3; 690,21; display, 647,24; set out, 663,14; *to eirêmenon*, make the point, 651,1

deixis, proof, 676,17; 691,4

dekhesthai, accept, 647,27-8; 650,20-1; receive, 649,19 *passim*; 650,22.24-5; 651,10.14-15.17; 654,16-17.20; 655,13; 657,12; 659,33; 662,16.20; 681,12.14; 688,14.19.20-1.23; 689,10.13.15.17; allow, 683,29; be occupied, 684,16; take in, 651,21

dektikos, receptive, 645,30-1; 646,24-5.29; 647,3.8; 655,10.20; 657,23; 658,6; 664,9; 684,15; 688,18; 690,2; 692,18; capable of receiving, 654,32

dêlon, clear, 645,24; 646,27.32; 649,32; 650,4.18; 651,6; 653,8.27; 654,4.19; 655,13.33; 656,23; 657,33; 661,1; 662,27; 663,7.32; 664,27; 671,14; 673,17; 681,3; 685,26; 686,27; 687,4-5; 690,23; 692,11.35; 694,21; clearly, 648,14; 651,19; 670,15; 672,19.21.34; 673,7; 674,32; 676,27; evident, 667,11.35; 669,23; manifestly, 685,7-8

dêloun, make clear, 646,22; 690,3; 694,16; imply, 648,8.10; show, 654,13; 677,7

deuteron, second, 648,29; 657,2; 659,31; 660,15; 668,1

diadosis, transmission, 661,12; addition, 661,13

diakena, voids of, 694,16,

diairein, divide, 649,28; 672,18-19; 678,26-7; 679,9.20-1; 680,2; 684,13.17; make distinctions, 661,4

diairesis, division, 667,13; 680,7-8

diairetikos, disjunctive, 649,8

diairetos, divisible, 645,28

diaitan, examine, 653,5

diakomidê, transmission, 655,33

diakônein, serve, 655,31

diakoptein, divide, 691,31

diakrinein, distinguish, 650,10; determine, 658,21

diakrisis, distinction, 652,9

diakritikos, which distinguishes, 652,8

dialambanein, separate, 652,1; 657,23; 683,19; 693,12

dialegein, discuss, 645,29; 670,1

dialeipein, be intermitted, 689,30-1

dialuein, solve, 662,7; dissolve, 680,26; destroy, 666,15

diamakhesthai, battle, 647,18

diapherein, differ, 646,27; 650,29; 651,16; 656,8; 665,6-9; 681,36; be different, 653,12; 682,14; (*to*) difference, 667,28; 671,28

diaphora, difference, 664,11; 666,14; 667,25.29; 670,9.11; 671,2.36; 677,19.27.33; 678,11.13.28; 679,4.19; 680,4; 690,13.28; 693,1.32; 694,11; different, 667,20

diaphoron, in addition to, 655,6; different, 657,6; 658,17; 665,8; 672,32; 678,15; 693,32

diarrhiptein, scatter about, 681,27,

diasaphein, make clear, 654,3

diaspan, tear in pieces, 664,13; tear apart, 664,20; 671,10

diaspeirein, intersperse, 684,16.18

diastasis, dimension, 655,24.26; 658,11; separation, 692,9; *diastatos*, dimensional, 657,37; 665,34

diastêma, interval, 645,30; 646,19.28.35; 647,33; 648,2.4.7.11; 649,14.23; 650,19.27; 653,2; 654,1 *passim*; 655,14 *passim*; 656,11.14.19;

657,8.10-17.19.25.29.33; 658,1-9.22; 659,9-10; 663,16; 664,33-4; 665,1.8.11-12.30-1; 666,5.9-11; 667,18; 675,21; 678,3.5-6; 680,17; 681,18.20.26-9.31.35; 682,1.3.4-5.13.15-17; 684,15.23; 693,8.21.31.34-5; 694,2.22.25.28
diastolê, expansion, 690,29
didaskein, give instructions, 661,8; teach, 688,6; *didaskalia*, teaching, 680,25
didonai, concede, 677,26
diekpiptein, make its way through, 693,14-15.18
diekptôsis, move down through, 693,28
dielenkhein, show the error, 647,16; examine, 680,31; prove, 681,9
dierkhesthai, traverse, 671,29; 676,1.5; 678,6
diestôs, dimensional, 658,9; 681,19
diêthein, filter, 662,29
diexerkhesthai, traverse, 674,27; 677,4
diexienai, traverse, 675,14-15.20; 676,36
diienai, be distributed, 651,3; permeate, 681,17.20; pass through, 660,32; travel, 661,27; 674,25; 675,1.25.28; 676,18.27.34
diistanai, isolate, 649,13
dikaios, just, 652,11
dikhôs, in two ways, 668,13
diodeuein, travel, 660,16
diodos, passage way, 693,22
diorizein, distinguish, 646,14; 651,24.30; 652,1.3-4; 653,21; 656,2; 693,4; determine, 653,8; define, 657,4; *diorisis*, distinguishing, 651,32
diorthôsis, correction, 692,17
diplos, double, 649,24
dizêsis, (Parmenides) enquiry, 650,13
dokein, seem, 645,27; 646,6; 648,26; 650,30; 651,8; 653,23.24; 655,25; 657,6; 658,26.34; 659,20; 662,23; 663,6.10-12.33; 665,25; 666,15; 667,30; 668,33; 670,22; 673,13; 680,25-6; 684,25; 685,3; 690,30; is seen, 649,12; purport, 659,31; (*to*) opinion, 653,25
doxa, opinion, 646,34; 651,25; 663,18; 668,33; view, 646,36; 648,15; 682,25; belief, 655,1.3; (common) use, 653,24
doxazein, express a view, 648,23; (pass.), be a doctrine, 649,7
duas, dyad, 652,5.14
dunamei, potentially, 663,23; 688,11.25.29.34; 689,36; 690,1.2.4; 692,2.7-8.12; 693,8-9
dunamis, power, 652,3; 663,6; 669,5.8; 677,33; 678,1; 693,14.20
dunasthai, be possible, 649,23; 660,3; 676,13; be potentially, 688,34; can, 649,27; 654,3; 657,25; 665,24; 666,18-19; 667,11; 668,1.5.12; 669,16.34; 683,18; be able, 664,20; 693,14; may, 658,1; 659,7; 676,12
dunatos, can, 653,10-11; 693,29; possible, 658,17; 659,11; 661,11; 662,2; 663,26; 670,16; 671,4; 687,17; 693,19.26; available, 692,16
dusdiairetos, hard to divide, 677,29
dusexôthêtos, hard to penetrate, 677,29
duskolia, be unhappy, 650,23
duskolos, hard, 647,25
duspatheia, resistance, 692,27

ean, permit, 648,12; allow, 694,6,
eidos, form, 650,10; 656,28; 661,3.5; 688,21; 689,30.32; 690,8.12-13; 691,20; 692,22.35.37; *eidikos*, formal, 690,13
eikein, yield, 671,2
eikotôs, reasonably, 652,31; 653,4-5; 675,25; 676,21; 688,6; *metabas*, by natural transition, 664,6
eiôthos, customary, 762,30
eipein, discuss, 645,23.26; say, 646,13; 649,26; 650,32; 652,31; 653,28; 654,19; 655,17.28; 658,13.18; 665,21; 669,32; 670,4; 673,26; 676,11; 677,31; 687,21.35; 691,9; call, 658,4; 690,9; state,

654,30; speak of, 654,10; *houtos eipein*, use the words ..., 646,5
eirgein, guard, 650,13
eisagein, introduce, 648,28; 649,35; 661,22.26; 662,14; 665,1; 679,4-5; 680,12; 683,34; 684,13.33; 687,13; 689,14; bring about, 692,6; give admittance, 648,33; *eisagôgê*, inclusion, 658,13
eiserkhesthai, enter, 647,30
eisienai, enter, 648,32; 689,14; come in, 660,11.18.30; *eisodos*, entrance 648,30; 651,6
eispnein, inhale, 651,27
ekei, THERE, 650,8.10; 652,8.10.12.16
ekhein, have, 646,24; 648,3; 650,10; 653,7; 655,24.26; 656,28; 567,20-2; 658,25; 660,2; 661,35; 664,9; 667,22; 669,5; 671,22.27; 674,21; 675,7-11.24; 677,3.32; 678,35; 681,10.18; 682,24; 685,30; 688,20; 692,11; 694,8; contain, 647,4.27; 648,5; 650,29; 654,14.22.23; 657,22.27; 662,21; 677,23; 694,10; be, 647,8; 653,6; 654,3; *homoiôs ekhon*, is similarly related, 666,23.27
ekluesthai, weaken, 668,30-1; 669,7-8; be overcome, 669,13
ekmuzan, suck out, 647,28
ekpiptein, fall out, 668,34; escape, 694,19
ekpheugein, avoid, 659,23; escape, 660,31
ekpurênizein, squeeze out, 660,3.9
ekroia, outflow, 659,25
ektasis, increase, 690,20
ekteinesthai, extend, 648,25; increase, 690,11; expand, 650,18
ekthesis, exposition, 672,31
ekthlibein, extrude, 660,3
ektithenai, set out, 652,29; 657,6; 683,5; expound, 653,18
ektropê, diversion, 667,15
elenkhein, refute, 647,16; 656,31; 657,37; argue against, 647,32; prove 648,1; disprove, 686,28; attack, 658,30
emballesthai, be put into, 650,22; throw, 659,24; 681,4; pour into, 663,2
emphainesthai, appears, 652,3
empheresthai, be carried about in, 657,27
empiptein, enter, 658,16
empodizein, hinder, 662,4; do not permit, 667,27; obstruct, 680,8; impede, 672,37-8; 673,3; stop, 670,5.13; prevent, 670,24-6; 693,30; *heauton empodizei*, self-defeating, 661,24-5
empoiein, introduce, 662,6
empsukhos, animate, 660,29
enankhos, recently, 675,7
enantios, opposite, 656,34; 667,2; 688,8.11-12.14.19; 689,1.15; 690,2.33; *enantiôsis*, opposite, 688,9.14-15.22
enargeia, fact, 661,31-2
enarges, clear, 688,20-1; evident, 670,24; plainly, 676,31; obvious, 650,1; (*to*) obviousness, 668,16.21; *enargôs*, obviously, 649,31; 675,13; evident, 674,32; plainly, 663,30
endekhesthai, possible, 659,24; 669,17; 682,20; can be, 662,26
endeiknusthai, show, 651,8; 669,18; 687,30; 691,20; point out, 658,13; indicate, 676,19
endidonai, give, 664,11
endoxon, common opinion, 651,8; doctrine, 651,25
eneinai, be in, 665,19.22; be natural, 685,18
energeia, actuality, 692,13; *kat' energeian*, actual, 694,6; *energeiāi*, actually, 648,10-11.17-18; 688,10.26.34.36; 689,36; 690,1; 691,26.28.30; 692,2.8.11; 693,9
engignesthai, come to be in, 656,28
enisthasthai, attack, 648,30; 661,6
enkeisthai, be inside, 647,25; be contained, 681,11
enkerannusthai, be compounded, 692,3
enkhein, pour in, 662,15; 681,7

ennoein, conceive, 646,1; realise, 648,12
ennoia, conception, 647,18; 648,3, 6; 653,7.13; 656,6.28.34; 657,6; 694,29; concept, 648,31; 649,5; 653,11.19.36; 654,6; idea, 676,25; notion, 670,31
enoran, see in, 656,10
enspeireisthai, be interspersed in, 684,24
enstasis, objection, 661,1
entautha, HERE, 652,12-13.17
enulos, enmattered, 690,9-10; 693,23
enuparkhein, be contained, 653,36; 660,1
eoike, seems, 647,3; 655,12.38; 666,34
epagein, add, 645,23; 649,6; 653,25; 654,30; 656,7; 662,5; 665,22; 666,9; 669,18; 670,17; 673,28; 674,21; 676,28.35; 683,6; 687,16; allege, 677,25.26; conclude, 650,3; point out, 646,22; 685,5; bring forward, 685,9; introduce, 682,27; adduce, 654,11; 656,33; 672,2; set out, 658,14; give, 655,27; object, 684,30; argue, 660,31
epangelleisthai, announce, 688,7
epekhein, stop, 659,24
epeiserkhesthai, be added, 688,24; replace, 689,29
epeisienai, enter, 651,27; 681,1
epekteinein, spread out, 683,13; extend, 689,34-5
epharmozein, apply to, 655,14
ephexês, next, 645,26; 655,12; following, 648,27; and so on, 649,25; from now on, 654,7; what follows, 654,13; *ta*, series, 651,31; members of a series, 651,32
ephistanai, maintain, 646,4.23; say, 686,20
epideiknunai, display, 647,15; exhibit, 647,27; 676,24; 687,31-2
epididonai, acquire, 661,12; increase, 661,19; 688,2; 689,8
epidosis, increase, 661,11; 683,14
epikatalambanein, catch up on, 679,24
(***to***) ***epikeimenon***, content, 685,2
epikhein, pour in, 662,28
epikheirein, try, 647,15; 655,3; 662,2.24
epikheirêma, argument, 647,15; 648,27.35; 649,4; 651,11.25; 652,18.28; 664,15; 665,29; 666,21.34; 670,18.22; 671,4.15.20; 673,21.30; 680,29; 682,12; *epikheirêsis*, argument, 651,8; 667,4
epikratein, overcome, 668,31; 669,8
epilambanein, cover, 679,8.10; receive, 693,34
epilêpsis, expansion, 687,21
epinoia, conception, 648,10; concept, 688,13
epiphaneia, surface, 665,5
epipheresthai, bear upon, 646,20; adduce, 654,21-2; add, 665,22.35
epipolazein, rest on the surface, 679,6; be above, 685,22-3
epipômatizein, stop, 647,29
epirrhein, run in, 662,9-10; flow on, 669,11
epispasthai, draw out, 652,23; drag, 685,12
episkeptesthai, examine, 646,31
epistasis, consideration, 660,26
epistasthai, note, 666,33; 687,11; 690,11; 692,6-7; know, 675,22
episurrhein, flow on, 669,6
epitêdeios, suitable in nature, 647,8; *epitêdeiotês*, suitability, 665,9; 690,20; 693,31
epiteinein, intensify, 679,28; 689,32; 690,9-10
epiteleisthai, be achieved, 661,28
epôthein, push on, 668,29; push away, 669,3
erein, say, 648,16; 657,26; 568,12; 659,27; 663,15; 667,31; 671,2; 676,15.36; 679,34; 685,10; set out, 652,32; state, 654,10; 657,29; 667,4; (perf.) above-mentioned, 657,19
êremein, be at rest, 670,13; rest, 670,13-14; 679,23
erkhesthai, approach, 648,31
erôtan, rely on, 650,4; put forward, 658,26; claim, 659,2-3; argue, 658,27

erôtêsis, postulate, 659,6; question, 660,31; required conclusion, 667,35
esterêmenon, deprived, 645,29, 646,17.25; 657,11; 658,23; 664,1; 693,8; 694,2,
ethein, be his custom, 647,14
ethos, custom, 646,35
eudiairetos, easily divided, 677,36; ease of division, 680,5
eueiktos, yielding, 678,12
eukinêtos, mobile, 668,28; 669,4.9; easily mobile, 694,20; easily moved, 669,1
eukolia, complacency, 654,6
eulogos, satisfactorily, 657,30; reasonably, 665,22
eupatheia, susceptibility, 692,27
eupathes, susceptible, 692,28
euthus, straight, 689,23.25.29; 690,22; 692,26; 694,25
eutheôs, readily, 647,28
exakontizein, shoot off, 660,10
exêgeisthai, explain, 656,7; 676,7-8; 678,31; expound, 668,33; interpret, 684,3-4; *hoi exêgoumenoi*, commentators, 676,16; *exêgesis*, interpretation, 655,12; *exêgêtês*, commentator, 670,17; 686,26
exienai, go out, 647,30; 669,1; 694,19; depart, 660,5; 689,14; come out, 681,8
existanai, be transferred, 650,8; make way, 680,34; 681,21; (*to*) departure, 688,30
exodos, exit, 681,9.10
exomoioun, assimilate, 662,12
exôthein, thrust away, 677,24; push aside, 677,28; push out, 686,34

gê, earth, 663,32; 664,29; 666,25; 670,1.28; 681,1
genesis, coming to be, 660,24; 679,21-2; 683,10.36; 687,13
genos, genus, 653,21; kind, 694,4
gignesthai, + gen., share, 648,15; + *ek*, come from, 660,21; 686,24.24-5; 688,26-7.30; + *en*, come into, 649,30; come to be, 691,27; be in, 649,30-1; become, 660,19.21; 677,30; occur, 649,9-10.18.21; 650,32, 651,2-4; 655,31-2; 678,19; be, 649,17; 661,35; come about, 651,6; 652,13; 661,2.4.21
grammê, line, 653,34; 672,23
graphein, write, 648,17; 678,30; (pass.) reading, 654,12.14.34; *graphê*, reading, 648,22; 655,22.26

hama, at the same time, 650,31; *einai* coincide, 651,5; at once, 666,8; simultaneously, 664,13.20
haphê, touch, 654,12-656,1 *passim*; 657,9
haplôs, unqualified, 646,35; independently, 691,28; simply, 658,2-3; 659,3; absolutely, 685,13; in general, 679,14
haptesthai, touch, 668,8
haptos, tangible, 653,3.30; 654,2.20.27.30.32; 655,2.13.20; 656,26
harmozein, suit, 664,34; 665,1; accord, 667,1; fit, 685,21; be apposite, 680,17
harpax, syphon, 647,27
harpazein, take up, 647,28
en' heautôi, in itself, 648,25; *kath' heauto*, in its nature, 648,24
kath' hauto, independently, 648,15; in itself, 651,19; 656,27; on its own, 657,22.24-5; 663,16; 682,21.27
hedra, seat, 669,2; 694,22; place, 694,21,
hêgeisthai, think, 665,11; 676,14
hêgoumenon, antecedent, 649,21
hêkein, come to, 658,14
hêi, qua, 646,29-30
hêlios, sun, 693,15.24
helkein, attract, 652,22; 663,7; draw out, 652,24; drag, 668,14; pull, 669,27-8.31
helktikos, attractive, 663,6
helxis, attraction, 663,4.5, 6
henôsis, unification, 652,12
hepesthai, follow, 653,34; 654,10; 665,20; 672,1; 674,1; 677,10;

679,34; 681,33; 685,9; 687,8; 691,4.10; result, 678,18
hepomenon, consequent, 666,6; 675,3; 677,18; what follows, 673,27.33
hermêneuein, mean, 686,30; *hermêneia*, interpretation, 665,18
heteroiôsis, becoming other, 691,25; becoming different, 692,28
heterotês, diversity, 650,9-11; otherness, 652,8.16
heuriskein, find, 672,8
hexei, positively, 647,5
hexis, possession, 647,3; holding power, 671,9.11-12
histasthai, come to rest, 670,4.10.23; be stationary, 670,15.17.25; stop, 670,27-8
historein, report, 663,3
hodos, way, 650,13
holkê, attraction, 652,21
holos, whole, 653,23; 659,19.21; 660,8; 662,31; 665,35.36; 666,1.3; *to holon*, the whole, 649,25.30, 664,18.23; 666,2-4; 671,11; 689,32
holôs, altogether, 646,7; 666,16; at all, 649,23; 657,18; 672,30; 677,22; generally, 660,6; 661,7; 676,30; 689,30; in general, 661,16.20; 665,23; 684,23; absolutely, 666,19; 692,15; totally, 685,28
holoteles, perfect, 650,6
holotês, the universe, 659,19
homoeides, homogeneous, 664,16; of the same species, 671,25
homogenes, of the same kind, 680,6
homoios, undifferentiated, 666,26
homoiotês, likeness, 664,23.32-3; similarity, 675,23
homologein, accept, 663,23; *to homologeisthai*, agreement, 653,19
horan, see, 672,1; 678,20; 682,2; 691,14
hôrismenon, given, 672,11; determinate, 672,34
horismos, definition, 655,25
horizein, define, 655,19; determinate, 690,16
horman, undertake, 647,13; start from, 649,8
hormathos, chain, 652,25
hormê, effort, 669,30
hudôr, water, 647,28.30; 649,20; 651,10.15.18.20-1; 659,13.24.26.30; 660,6.12.21-2.24-5; 661,19; 662,15 *passim*; 663,2; 665,10; 670,33-4; 671,28; 673,1-2.4; 674,12; 679,6.8; 680,34; 681,4-5.7.21; 682,37-8; 684,2.5.35; 686,24-8.31-2; 687,1.3-4; 688,23-5.27.29; 693,14.27
hudraules, hydraulic organ, 681,7
hugros, liquid, 658,21; damp, 669,1; moist, 694,20; wet, 687,5; 688,9; 693,16
hulê, matter, 646,11,15; 656,27.29-35; 657,2; 658,2; 661,3.5; 687,34; 688,5 *passim*; 689,3.7.13.15.34; 690,3 *passim*; 691,15; 692,18.30.32.37-8; 693,1.9.25
hupantan, respond, 645,31-3; confront, 659,17; deal with, 661,7; meet, 663,18; reply, 661,9-10; *pros ti*, oppose, 647,16; 648,5; attack, 684,13
huparkhein, be present, 648,12; hold of, 657,14; be, 664,12; 671,34; 678,28; 679,7; exist, 663,34
huparkhon, attribute, 681,33; (plur.) character, 653,13.15
huparxis, state, 667,16
hupeikein, yield, 671,1.8
to hupeiktikon, yielding nature, 670,32
hupekrein, run out, 662,8
huperairein, exceed, 685,37
huperballein, exceed, 672,9; 673,29
huperekhein, exceed, 672,15-18.23.25-6; 675,30; 676,4-6.12.14; 685,36; 686,10-11
hupêretein, serve, 655,33
huperkhein, overflow, 650,25; 684,6; pour over, 683,31
exô huperkheisthai, stick out, 687,16
huperkhusis, pouring over, 683,32
huperokhê, excess, 671,33; 672,18.20.26; 678,8; 685,32; 686,1

hupexistanai, yield, 681,15.17
huphistanai, exist, 648,15; subsist, 693,22; be, 668,12
huphizêsis, settling, 685,23
hupoballein, suppose, 648,7
hupodeiknunai, indicate, 645,33-646,1; suggest, 657,7; point out, 687,30
hupodektikon, receptive, 657,31-2
hupokeimenon, substrate, 646,26-7; 649,28.30; 656,31; 674,16.19; 677,16; 678,18.26; 679,9.11; 680,8; 688,5.8.15; 689,2.26; 690,2.24; 692,14; 693,21; burden*, 685,1
hupokeisthai, underlie, 692,39
hupokhôresis, yielding of place, 681,6
hupomenein, be permanent, 665,28.33
hupomimnêskein, recall, 680,16; remind, 691,17; 692,17
huponoia, doubt, 659,23
hupostasis, existence, 646,8; 688,11; 693,21.30; substance, 652,6; foundation, 663,14; *en hupostasei einai*, exist, 648,9-10; be substance, 692,21
hupothesis, hypothesis, 657,28; 661,29; 677,10; 683,36; 684,17.25; 693,19; posit, 674,26; *ex hupotheseôs*, *ex hypothesi*, 676,17
hupothetikos, hypothetical argument, 663,22; *hupothetikôs*, hypothetically, 663,26
hupotithenai, form hypothesis, 654,6; hypothesise, 654,7; suppose, 654,23; 661,13; 665,34; 666,9.11; posit, 656,24, 658,11; 661,24.33-4; 662,1.32-3; 663,9.17-18; 670,12.31; 673,33; 674,5; 676,21.30; 677,9.14.22; 680,34; 681,34; assume, 671,37
husteros, posterior, 647,4; 667,16; later, 648,16

idein, observe, 654,26; see, 692,35; know, 686,28; 694,13
idios, particular, 647,3; *idiotês*, special nature, 665,5
ikhthus, fish, 659,27
iskhuros, powerful, 655,18
iskhus, strong, 670,13; 676,15
isomegethes, of equal size, 678,22
isometros, of the same size, 694,23,
isorropos, equipoised, 666,26.32
isos, equal, 649,2-9.32-3; 666,5; 684,4-5; 686,24.26.28-9.32; 687,16.26; 688,2; same, 685,36; *hoson*, as much as, 651,10; *isotês*, equality, 675,22; 683,38; 686,30.33
isotakhes, at the same speed, 677,26; 678,9; *isotakhôs*, at the same speed, 678,27; 679,17; at equal speeds, 679,24-5

kaiein, burn, 662,28-9
kairon ekhein, time has come, 653,5
kakôs, wrongly, 657,37
(***to***) ***kalon***, beautiful, 652,10-11; *kalôs*, well, 646,4.23; 676,8; 677,11; 686,20; 687,25.35; *mê kalôs*, badly, 647,17
kaloumenon, so-called, 668,1
kanôn, criterion, 672,17
katalambanein, catch up with, 679,17-18; understand, 681,6; take up, 683,15-16
kataleipein, neglect, 675,3
katallôlos, appropriate, 647,23; consistent, 655,28; (adv.) consistently, 665,15
katapheresthai, fall, 668,32; 678,22.31; travel downwards, 679,3
kataphronein, despise, 654,5-6
kataskeuasma, apparatus, 681,6
kataskeuazein, confirm, 660,15; support, 663,27; establish, 665,15.25; 678,2; 680,26; *kataskeuê*, argument, 661,6; structure, 683,5; *kataskeuastikon*, in support of, 651,11; in favour of, 657,31
katastrephein, turn upside down, 659,24
katekhein, stay, 659,12; maintain place, 659,14; 687,21; occupy, 666,5; 681,30.35-6
katho, ***qua***, 647,7-8
katholou, universal, 645,25
kathoson, ***qua***, 646,25

katô, below, 664,12; 665,9; 667,21-2.28; downwards, 669,32; 679,27; 684,30; 685,11-12.14.17-18.32; 690,26; bottom, 693,18; *epi to*, downwards, 663,32; 669,9; 671,26; 678,20; 687,26
katôphoron, moving downwards, 671,32; 685,22; travel downwards, 685,18
keisthai, lie, 666,22
kenos, void, *passim*; empty, 651,10; 662,16; 694,5.17; *kenotês*, void, 694,14; *kenôma*, empty space, 651,15-16; void, 662,26
kharaktêrizein, characterise, 656,10; *kharakter*, character, 691,4
kharientôs, appealingly, 684,35
khôluein, be lame, 685,20
khrasthai, use, 646,35-6; 650,1; 659,16; 664,15; 667,8; 671,5; 685,3
khôra, room, 646,7; 683,11; place, 650,9.11; 681,36; 693,27.37; space, 683,14-15.28; 694,5.9.17.22
khôrein, pervade, 648,24; 649,19; occupy, 649,24; 659,16; pass into, 662,17; 681,21.28; permeate, 651,5; 682,7; spread through, 682,22; move through, 680,13; pass through, 660,17; 662,17.22; 680,32; travel through, 661,30.33; (pass.) be contained, 649,20
khôrêma, receptacle, 561,17; *khôrêtikon*, receptive, 693,31
khôrismos, separation, 651,31; 652,10.13
khôristos, separate, 648,8-9.17-18,20-3; 649,5; 656,5.35; 657,1.32; 658,23; 665,11.30; 684,25; 694,27; *mê*, unseparated, 684,16
khôrizei, separate, 652,2.15; 657,13.20.26; 659,9; 663,9.15.19.21; 664,34; 665,27; 666,2.20; 680,17; 681,29.34; 682,13.21; 683,3.33; 684,14.20.23.33
khreia, need, 651,28; 661,29; 691,3; 694,7; use, 676,8; *parekhesthai*, fulfil a need, 651,28; fulfil useful purpose, 682,8-9
khrôma, colour, 654,22.27.33; 655,5; 656,13
khronos, time, 645,23; 671,24; 672,5 *passim*; 673-7 *passim*; 679,2; 685,35-6; 686,2
khrusos, gold, 671,35; 678,16.21-2
kinein, raise, 661,14; cause motion, 684,34; move, 669,12.23.31; 670,2
kineisthai, change, 646,10-11.18; 659,8.12.15.18; motion occurs, 658,19; move, 650,2-3; 659,13; 664,5-7.19-20.25; 665,21.24-5; 666,7.22-3.33; 667,24; 668,8.17-18.26.29; 669,30.33-4; 670,3.4.10.12-13.19.23-4.27; 671,14.24.26.31.35-6; 672,4-5.10-11.32.34-6.39; 673,3; 674,12.14-15.27-31; 675,1; 676,29-30.32.35; 677,9.17.21.24.27.32; 678,3.6.13.18; 679,24-5.30.33.36.37; 680,1.18; 682,14-15; 683,12.29; 685,32.34.37; 686,2.5.13; 687,14-15.19-20; 690,23.25; 694,24.26-7; be moved, 669,23; *topikôs*, change place, 659,15; 683,25; (*to*) motion, 669,10.24; 670,33; 671,3; 679,23; 684,21
kinêsis, change, 645,22.24.26; 648,27; 649,8.12-13.17-18.21.35; 650,32-3; 651,2; 652,21; 657,32; 658,15.18; 659,3-4.7.21; 661,8-10; 663,10 *passim*; 666,18.20.33; 682,27; 683,8.20.23; 684,6; 685,13; 687,11.23.24-6; 691,15; 692,17.30.37.39; 693,2; motion, 658,27-30; 664,11.13.26-7; 666,16; 667,6 *passim*; 668,2-7.10.13.20.30; 669,3.7.35; 670,6-7.18.20; 671,13.17.19.21.25.27.37; 672,2.12.28-30.33.37; 673,5-7.11-12.16.21-2.24-6.34; 674,3-4.7.17-18; 676,2.5.33; 677,7.11.13-16.23.25.30.34.36; 678,2.4.10-11.19.23.33-4; 679,26-30.35.38; 680,2-3.12.14-15.26.29.30; 683,3;

684,27.30-3; 685,5.10.14-16.30; 686,12.14-15.19.23; 687,17; 688,1; 690,22.27; 692,16.26; 693,9
klepsudra, water-taker, 647,26
klinein, incline, 666,30
koinos, common, 645,25; 653,24; 661,25; 662,2.4; general, 647,1-2; 661,5; *koinê ennoia*, common conception, 646,33.35; *koinônia*, kinship, 689,19
kôluein, prevent, 646,15; 659,1; 667,23; 669,30; 681,22; 682,6; 693,26; *kôlutikon*, prevent, 664,27
kosmos, universe, 648,13-14.21; 671,6-7; 691,30; cosmos, 648,24-5; 651,27; 652,8-9.15-16
kouphizein, make light, 685,1.4
kouphos, light, 653,32.35; 655,35.37; 671,34; 678,32; 684,28; 685,35.37; 686,2.6-8; 690,18.23.25; 691,5.9.11.13.21; 692,24.26.33; 693,2; (*to*), *kouphon*, light thing, 650,29; 685,19; *kouphotês*, lightness, 653,31; 655,2; 656,12-13.15; 671,34; 678,20; 685,19-20.33-4.37; 686,1; 692,4.19
kreittôn, more powerful, 670,5.24
kritêrion, criterion, 646,35
kuathiaios, a cup of, 649,20; 683,37-8; 684,2
kubos, cube, 680,34; 681,4-5.17.21.33.34; 682,1
kuklophorêtikon, in circular motion, 655,36;
kuklos, circle, 689,19.21; 690,28; 691,1; 694,25; *en kuklôi*, circular, 683,28; 687,18-20.24; *kukloteres*, round, 694,4,
kumainein, bulge, 683,23.35; 686,19.23.35; 687,15.19.26; 688,1
kurios, important, 658,14; literal, 661,24; proper, 661,13.21; ***kuriôs***, true, 682,23; in primary sense, 686,5
kurtos, bent, 689,21-3.25.27.29; curved, 690,32; *kurtotês*, curvature, 689,25; 691,1-2

lambanein, settle, 653,1; gather, 653,13; take, 665,5; 672,20; 674,7; 675,15; 676,22; posit, 672,16; receive, 669,10; pass., be taken, 654,13; 655,27
legein, say, 645,31; 646,10.14.17.19.23.31; 647,12-13.20.24; 648,2.11.13.21.23.33; 649,8; 650,2; 651,1.26; 652,19.30; 653,23.26.29-30.32; 654,28.30.36; 655,15-16.23.34; 656,2.6.9.11.23-4.28-9.30-1.33; 657,36; 658,5.26.34; 659,2.18.20; 660,1.7.27; 661,9.15; 662,7.19.25.33; 663,9.18-19; 664,22.34; 665,4.12.17; 666,13; 670,13.27.34; 671,4-5.8; 672,2.22; 673,34; 675,14; 676,16; 677,10.22; 678,19.35; 679,7.13.28; 680,21.28; 681,25; 682,22.29; 684,4.19.33; 685,9.17; 686,6-7.18.26.34; 687,22; 689,14.28.35; 690,4.30; 691,1.5.26.29; 692,3.8; affirm, 646,1; 649,4; 652,28; 653,4; 656,35; 657,21; deny, 653,4; call, 653,35; 654,24.33; 658,2.19; 670,19; 672,22; 684,15; 686,7; describe, 654,8.11; mean, 654,15.35; 661,11.14; speak, 646,33; 656,25; 657,22; use (the conception), 647,18
legesthai, call, 646,24; be called, 646,11; be applicable, 648,33
legetai, discussion, 676,10
leipesthai, be left, 694,6,
leptomeres, light, 660,5.20-1; with lighter parts, 672,1
leptos, light, 660,3-4.6; thin, 672,4.33-4.37-8; 673,1.2.18.26.28; 676,1-7.12.18-20.23.26-7; 678,12; 692,10; *leptotês*, thinness, 676,23-4
leptunein, become thinner, 688,33
leukainesthai, turn white, 658,32
leukos, white, 655,33; 681,23.30; 688,16.20.33; 690,14; bright, 690,5
lexis, sentence, 665,20.26; text, 670,21; discussion, 675,14; passage, 673,24
lithos, stone, 652,22; 663,7; 670,33.34; 681,29; *sidêritis lithos*,

iron-stone, 652,23.25; *lithinos*, stone, 680,34; 681,3.21
logos, discussion, 645,31.33; 646,13; 656,31; 666,6; 675,3; 678,3; 693,5-6; discussing, 669,35; position, 685,21; proportion, 686,1.4.11; treatise, 691,4; definition, 646,26; 681,29.682,13; 688,12; account, 646,37; 649,35; 650,4.17; argument, 652,29.31; 657,16.29; 658,24.30; 659,31; 660,1.19; 661,8.16.22.24; 662,14; 663,3.11; 664,30.34; 666,15; 667,8; 668,1; 671,18; 676,23; 680,18.26.28; 682,2; 683,1.2.5.33; 684,13; 685,29; 687,29; 688,7; 692,3.16; thought, 647,14; ratio, 671,22; 672,8-675,24 *passim*; 677,3.12.15.17.35; 678,35; 679,1; 685,30.35; relation, 675,6.8; *sumplêrotikos*, determining account, 690,13; *ekhein logon*, be reasonable, 648,22-3; *kata logon*, rationally, 665,33; *para logon*, irrationally, 665,35
luein, dissolve, 659,5; solve, 661,25-6; 662,2.23-4.34; 663,13; 688,2; 693,19.29; escape, 687,22; refute, 661,17; 683,3; answer, 660,32
lusis, solution, 654,29; 655,27

manos, rare, 657,23; 680,22; 683,31; 684,10-11.19.27-9.31; 686,6.10; 690,5.18.22.24-5; 691,6.8-9.11-12.19-20; 693,26; rarity, 682,19.28; *manotês*, rarity, 685,34; 691,32; 692,4.6.18
manôsis, rarefaction, 682,29; 683,4.6.8.17-18.22-3; 684,8; 686,18-19.21-2; 687,8.10.20; 688,4; 690,19; 692,7.32; 693,25.33
manousthai, become rarer, 689,12; be rarefied, 693,34-5
manthanein, be taught, 653,10; learn, 661,13
marturion, witness 651,9
matên, pointless, 663,34
mathêmatikon, mathematical, 658,1.3.4; 665,4; 690,10
to mêden, be nothing, 648,2; 674,22
megas, big, 649,19.22.27-9.31.33; great, 650,4; 684,18-19; 688,17
megethos, magnitude, 656,13; 674,18; 678,35; 690,12; size, 661,12.18; 674,11; 678,15; 679,16; 681,33; 688,2.18.21-2.33.36; 689,18.33-4; 690,16-17.21.33
meiôsis, diminution, 676,25
meiousthai, diminish, 662,31; decrease, 662,30
mêkos, length, 675,2
memphesthai, criticise, 658,33
menein, remain, 650,18; 650,7.9; 659,22; 668,30; 669,3; 681,17; 687,24; 688,13.22.26; 689,15-16.36; rest, 664,7.14.21.29; 665,9.15-16; 671,9.11; be at rest, 665,22.24-5; be static, 678,12; stay, 671,6; be stationary, 671,29
meros, portion, 661,31; part, 661,32; 662,8; 664,23; 665,28; 666,4.7.27; 670,14; 672,22.25; 675,18; 676,22; 689,20; 693,28; 694,14; *merikos*, partial, 650,5
mesolambanesthai, be separated, 648,20
mesos, middle, 663,22; 666,25-6; 667,21-2.28
mestos, full, 650,24; 681,7
metaballein, change, 660,20; become, 683,31; transform, 683,37-8; 684,1; 687,1.12.19-20; 688,9-10.25-6.30.34.36; 689,1-2.4; 690,34
metabolê, change, 660,24; 661,19; transformation, 684,4.7; 687,5.7.16.26; 688,2.23-4; 692,28.33
metakheirêsis, approach, 687,31
metakhôrein, take over, 694,17
metapherein, transfer, 682,14
metapiptein, change, 667,24
metastasis, exchange, 649,13
metaxu, between, 648,11; 651,32; 652,1.5.15; 678,3-4; *to m.*, medium, 649,9.11.14; middle, 681,1-2,; intermediate, 663,32
metekhein, participate, 652,9; 660,29; share, 665,6
metrein, measure, 678,6-7

metretikos, measure, 673,7
mignunai, mix, 663,17; 684,7; 686,8; 688,28; 692,4
mikros, small, 649,19.22.28-1
mixis, mixture, 688,29; 690,6; admixture, 689,6
molubdos, lead, 679,5; 691,8.17.22
monas, monad, 652,4.14
monê, rest, 664,13-14.28; 666,25; 670,8
morion, part, 659,14; 660,4; 664,17-18; 665,36; 666,2.23; 671,3.10; 675,6; 681,27; 690,4-5; portion, 674,31

nekhesthai, swim, 659,26
noein, think of, 657,12; conceive, 692,23; *noêma*, thought, 650
noêtos, intelligible, 650,5; conceptual, 690,10
nomizein, think, 647,19; 648,1; 655,17.21; 675,20

oiesthai, think, 646,21; 647,32; 648,22; 650,7; 656,20; 657,37; 662,32; 664,30; 666,18; 670,7.20; 676,9.28; 679,38; 682,19.26 28; 684,1.17; 687,30; 692,6; 693,19; believe, 651,16
oikeios, concern, 645,33; own, 669,30; 670,3.9.19; 678,1; 682,15; 690,16; innate, 679,32; proper, 648,30; 664,16-7.19; 665,16; 667,20
oinos, wine, 650,21-2.25-6; 651,12-18; 659,33
okhein, carry, 668,9.14; 669,18.20.25.28-9.31; 685,27
met' oligon, soon, 648,16; shortly, 684,30
kat' oligon, gradually, 668,30; 669,7.13; 676,24
on, is a reality, 651,11; is real, 650,11-12; *to on*, what exists, 648,1; what is, 650,2-3; 653,26.29; reality, 658,13-14; *to me on*, what is not, 650,9.12; 658,11; that is nothing, 678,6; which is not, 652,12-13; non-existent, 652,14-15; 667,30; not-being, 652,16-17
onkos, bulk, 651,18; 660,19.21; 681,3.4; 683,13; volume, 660,4.8; 661,20; 662,31; 663,1; 681,32; 682,4-5; 686,26; 687,1.3.7.12; 688,27.30.35; 689,34; 691,19; 694,20; quantity, 681,16; mass, 669,1
onkoun, increase, 693,22
onoma, word, 653,2.8.12.18.20.23; 657,3.17; 658,21; *epitrepein*, call, 682,20
onomazein, name, 672,39; call, 694,13
opê, hole, 647,29; 681,8
ôthein, press, 663,1; push, 668,14; 669,1.16; 687,1; 694,1.20
to ouden, nothing, 673,8; nought, 675,24
ouranos, heavens, 656,14; 659,19; universe, 657,23.26; 687,16; *ouranion*, heavenly, 656,15
ousia, substance, 650,18; 656,7.12.24; 657,9; 660,8; 662,32; 692,34.38

pakhumeres, dense, 660,7-8; with dense parts, 671,37
pakhunein, become dense, 681,12; 688,33
pakhus, dense, 660,6-7; 672,6.36-7; 673,1; 676,2.7; 677,29; 678,11
palaios, antiquity of, 649,35
to pan, the whole, 648,20.21; 658,33; 666,29; 690,29; universe, 683,24.33.35; 684,6; 686,19-20.23; 687,26; 688,1; 694,3.7
pantakhêi, throughout, 651,7; everywhere, 671,8
pantakhou, everywhere, 647,4; 648,25; 658,10; 665,8; 692,12; universally, 667,16; in all directions, 671,2-3; *pantakhothen*, on all sides, 651,7; 666,29; from all sides, 670,31
pantelôs, wholly, 650,9.11; *to pantelôs on*, which wholly is, 652,12
pantos, altogether, 660,17; certainly, 667,14; always, 676,20; 683,35
paraballein, compare, 646,36
paradeigma, illustration, 651,16; 684,35; example, 659,16.22-3.32

paradeiknunai, compare, 656,34
paradekteon, acceptable, 686,20
paradidonai, hand down, 656,6; provide, 672,17
paragein, introduce, 647,17
parakolouthein, follow, 676,16
paralambanein, take together, 673,16-17; introduce, 680,27
paraleipein, omit, 663,25; 668,16.21; left, 694,14
parapherein, bring out, 647,24
paraphrasis, paraphrase, 691,2
paraplêsios, similar, 646,15.21
paraspeirein, intersperse, 663,17; 683,17; 687,9; 691,31; 692,1.5.8; scatter, 682,24
paratithenai, put forward, 650,20; add, 651,26; 684,35; set out, 652,28; 653,4
parêkein, omit, 649,34
parekhein, contain, 671,2; give, 694,29; excite, 676,25
pareinai, be present, 668,14-16; 669,13.27; 670,2
pareisdusis, intervention, 652,13
kata parêmploken, weaving in, 691,20
paristanai, support, 647,14; 684,12; make clear, 686,26-7; rejoin, 687,22
parodos, entry, 661,22; 692,11
paroimia, adage, 648,33; proverb, 661,23
parorama, oversight, 659,4
pas (***panta***), all things, 645,25
paskhein, be acted, 692,34
pathêtikos, susceptible to change, 693,25
pathos, condition, 682,13; feature, 681,24.36
peirasthai, try, 648,32; 661,15; 693,12
peithein, convince, 653,12
peperasmenos, limited, 645,27; 673,9.12; determinate, 673,22-3; 674,5.19.20; 677,12
pephukos, capable of, 654,15; naturally, 681,29; 692,13; 694,5; suitable, 654,18; be suited to, 654,27; be of a nature, 657,1; be natural, 670,14
pephusêmenos, blown up, 647,24.26; 685,1
peplêromenon, full, 646,29-30
perainesthai, end, 671,20
peras, limit, 657,15; 665,3-4.6-7; 667,21; end, 673,14
peridineisthai, revolve, 659,11
periekhein, contain, 657,15; 665,4-5; surround, 671,5; 692,10-11
perigraphê, boundary, 684,20
to perikekhumenon, which surrounds, 651,28
perikrouein, attack, 670,31
perilambanein, include, 662,23
periodos, circle, 694,4
periousia, superabundance, 689,21
peripheresthai, swirl round, 659,13-14
periphereia, periphery, 689,12.20.23.27; 690,28; 691,1
peripiptein, fall into, 656,26; 663,13; (pass.) be entrapped, 677,25
periplakein, enclose, 679,18
peritrepein, turn round, 661,14; 680,28; (pass.) *aporia*, the problem rebounds, 662,18
peritton, excessive, 650,23
to perix, surround, 667,21-2.28
phainesthai, appear, 653,7; be seen, 653,14; 691,22; 694,1.12; seem, 654,8; 680,23
phanai, say, 645,24; 646,2.11.16.30; 649,29; 650,14.19.26.29; 651,23.30; 653,5.8.27.30; 654,34; 655,14; 657,5; 658,14; 660,27.30-1; 661,1.28; 662,16.18; 663,16; 664,16; 665,31; 666,25; 667,29; 674,11; 676,10.14; 679,12; 684,29; 686,17; 692,2; call, 658,3; 668,16-17; 671,1; 673,14; affirm, 647,12-13; 652,26; *mê/ou*, deny, 647,13.17; 652,26; 686,16
phaneron, clear, 661,17; 663,5; 671,11; 677,7; 682,10.28; 691,29
pheresthai, move, 659,25; 679,9; travel, 664,11; 665,2.14-16; 668,9; 669,10.19; 670,16.19.30.27; 671,7; 673,29; 676,28; 677,8.19.28-9.36;

678,1.8.21.25.28.30.35; 679,1.15; 681,31; 685,4.6-7.11.19.20.27.33; motion, 667,25; 669,17; be carried away, 671,10-11; be carried along, 668,4; be found, 667,4
pheugein, avoid, 661,33; 662,18.23; 680,33
phora, local motion, 649,9; 650,32; 651,10; travel, 663,27.32; 665,7; 668,32; 669,9-10; 679,20-1.31; 685,5-6.24-5; 691,25; motion, 669,14
phôs, light, 693,14.20.23-4
phthesis, decay, 662,10
phthora, passing away, 660,25
phulattein, retain, 659,19; preserve, 676,20; 683,38; 687,11; keep, 678,17
phusikos, natural, 663,30; 665,2.5.7; 669,34-5; 670,2; *ho* natural scientist, 645,23.33; *phusiologountes*, naturalists, 653,26
phusis, nature, 651,24.30; 652,4; 663,24-5.2-9.33; 665,34; 678,33; 681,19; 684,34; 685,7.15.20; 689,15; 690,23; being, 653,27; *phusei*, naturally, 646,37; 685,22; *kata phusin*, natural, 663,31; 664, 27.28; 666,33; 667,2.6.9-10.12. 14-16.19.25-7.31. 33.34-5; 668.2. 3.4.6.10.11.17.19. 23.32; 669,23. 24.33; 670,6.8.15.17-19.25.26; 671,31-2; 679,31; 680,5; naturally, 664,18.23; 688,19.20; *para phusin*, unnatural, 667,2.3.10.13.14.17; 668,11; 670,7.25; contrary to nature, 670,3; 686,7
phutikê, vegetative, 660,29
phuton, plant, 690,15
pilein, compress, 651,14; 660,2; 681,11; 683,11.16; 694,1
pilêsis, compression, 648,28; 650,17.27-8; 651,12; 652,21; 659,32-3; 660,5; 681,12; 683,16.28.34; 684,8; 686,34; 687,25; 688,3; 694,15
pisteuein, confirm, 650,1; 666,24
pistis, belief, 646,16.18.20.22
pithanon, plausible, 647,14; 684,25; persuasive, 648,34; 687,29
pithos, jar, 650,20-1.24; 651,13; 659,32; 687,6
plêres, full, 646.24.26-8; 647,1-2.6-7; 651,4; 660,32; 662,13; 671,1.22-4; 672,8-677,35 *passim*; 678,25; 680,13; 685,30; filled, 648,8; 649,10.15-16.18.21; 654,11-12.14.17.26.34; 655,8.23.25.28; 657,8; 658,32; 662,34; 668,25; 669,14; *to*, content, 693,17
plêroun, be complement, 650,22; fill, 651,19; 654,16-17.27-8; 655,19; 657,25; 681,24; *peplêrômenon*, full, 659,23-4; *plêrôsis*, be filled, 647,9
plêsiazein, neighbour, 693,35; 694,1; *plêsios*, adjacent, 669,1; 694,20
plêthos, multiplicity, 648,13; crowd, 683,12
pneuma, breath, 651,27; air, 681,8
poiein, make, 646,18; 651,7; 664,34; 672,31.38; 686,35; 688,31; act, 692,34; state, 688,7; construct, 668,15; *marturion*, call to witness, 651,9; *aporôteron*, exacerbate, 661,26; make more intractable, 662,2-3
poiêtikon, efficient cause, 664,3-4
poion, what kind, 659,1; qualitative, 688,14; quality, 689,18
poiotês, quality, 654,22.24.33; 655,6.10-11.30; 680,5-6; 688,38; 689,1.16; 692,33.37
hoi polloi, generality, 647,18
polukenon, contains many voids, 670,34
porizein, provide, 668,12
poropoeisthai, become channel, 662,13
poros, pore, 651,20; 652,23; channel, 662,11.12; 692,10; duct, 693,16
poson, quantity, 658,1.3; 688,16-17; amount, 662,20; *posotês*, quantity, 689,18; 692,33.38
pragma, thing itself, 653,15; thing, 656,35; 666,26
pragmateia, subject, 669,34

pragmateiôdes, weighty, 658,26; complicated, 671,18
proagein, extend, 676,22
proairetikos, intentional, 669,35
proapodedeigmenon, prior demonstration, 667,18
proballein, raise, 646,9; form, 653,11; set out, 682,25
problêma, problem, 646,2; 653,6.9.17; 680,31
prodêlos, manifest, 649,15
proêgeisthai, be prior, 647,2; 653,9.16
proêgoumenos, prior, 646,37; primary, 667,16; direct, 668,12; *proêgoumenôs*, pre-eminently, 663,14; primarily, 668,22-3
proerkhesthai, proceed, 690,13-14
proienai, go on, 668,19
prokeisthai, face with, 657,17; be projected, 673,31; the projection is, 668,22; be envisaged, 676,31
prokheirizein, exhibit, 647,14; undertake, 645,25; confront, 656,32
prokheiros, at once, 679,37-8
prolambanein, start, 653,20; be already, 668,21-2
proôthein, push in front, 668,27.28; push forward, 669,5; 683,27; push out, 683,31
propherein, confront, 661,19
prophora, proposal, 655,28
prosagein, add, 671,18; base on, 651,13; *mê prosapagoreuein*, deny, 692,13
prosdeesthai, have need of, 658,33
proseiserkhesthai, enter, 688,28
(***to***) ***proseisienai***, addition, 688,32; 690,7
prosekhês, nearest, 676,23
proserkhesthai, be added, 662,8-9
prosgignesthai, be accretion, 662,8; be added, 689,11
prosienai, approach, 683,28
proskhôrein, approach, 666,28
proskrinomenon, accretion, 660,18-19
proskrisis, accretion, 661,2.21.27
proslambanein, add, 650,2; 655,9; 671,24-5; acquire, 689,34; gain, 655,5; qualify, 664,8; take over, 659,15
proslêpsis, additional premiss, 663,23.27
prosmignusthai, be mixed with, 689,11
prosphues, suitable, 659,22; *prosphuôs*, appositely, 684,35
prosphuesthai, grow into, 662,12
prospiptein, fall on, 693,28
prosthêkê, addendum, 676,10; increase, 679,26; addition, 690,31; 691,14
prostithenai, add, 652,21; 656,26; 660,16; 668,23; 675,29; 676,17; 677,11; 679,28; 680,4.21; 685,24; 686,5; 687,2; 689,26; 690,34; 691,9.19; 692,39; attach, 681,8
prostithesthai, support, 646,33
protasis, proposition, 663,25
proteron, until, 647,29; previous, 656,8.14; earlier on, 656,21; former, 648,22; 656,18; first, 653,8
protithenai, propose, 645,23; project, 647,12; 658,24; set out, 680,31; 684,13
prôtos, first, 646,31; 647,13.17; 652,3; 653,23; 659,5.17; 660,15; 667,12; 668,3; first sort, 648,28
proüparkhein, be before, 681,1; 689,11
proüpotithenai, propose, 694,5
psêphis, pebble, 659,24.25.26
psophos, sound, 654,22.24.33; 655,5; 681,9
psukhê, soul, 660,29
psukhros, cold, 688,8.15; 689,2; 690,15
puknos, dense, 672,33; 680,22; 684,12; 690,18.22.26; 691,5-7.9-10.12; 692,23-5; 693,28; density, 682,19; *puknotês*, density, 692,19
puknôsis, condensation, 662,33; 682,29; 683,4.6.8.22-3; 686,18-19.21; 687,8.10.20; 688,4; 690,19; 693,26.33; density, 692,30
puknousthai, be compressed, 693,27; be densified, 660,3; grow more dense, 683,13; be condensed,

659,24.29; 660,8; 683,30; become denser, 689,12
pur, fire, 660,6; 663,31; 664,29; 665,9; 668,34; 669,2; 681,1; 683,38; 684,29; 690,32; 694,19.21-2
purên, olive-stone, 660,9

rhaidion, easy, 647,6; 661,17; 677,20; 679,7.25; *rhaidiôs*, easily, 667,24; 679,9
rheptein, have impulse, 679,11
rhêsis, passage, 690,30
rhêton, passage, 662,6
rhiptein, throw, 668,15.26-7.30; 669,3-5.8.13.26.32; 670,1.2; *rhiptomenon*, projectile, 668,27.28.32; 669,5-6.10-12.14.25; *rhipsis*, throwing, 668,25-6; *rhipsis gignomenê*, be thrown, 668,26; 669,19
rhopê, impulse, 655,2; 671,31; 677,32-3; 678,15.17.20.28; 679,1.30.32; 680,4-8; tendency, 663,31
rhumê, impulsion, 668,27; impulse, 668,31

salpinx, trumpet, 681,8
saphes, obvious, 663,25; clear, 672,38; *saphôs*, clearly, 657,28; 660,28; 676,24
selênê, moon, 693,23
sêmainei, mean, 646,4; 653,1.8.12.18.20.23; 657,17.20; designate, 657,3; show, 673,24; *sêmainomenon*, what is signified, 653,14; meaning, 653,20-2; 658,21
sêmasia, meaning, 654,10; 656,8
sêmeioun, mention, 667,3; *semêion*, point, 653,34; 654,4
sideros, iron, 652,23-4; 679,5; 691,8.18.22; *siderion*, iron thing, 652,22; piece of iron, 652,25
skedannunai, scatter, 671,9-10
skeptein, examine, 645,25
skhêma, shape, 656,13; 660,10; 671,27.32; 678,14-15; 679,4.9-10.17.19.20; 689,17; figure, 657,2; 690,32; *skhêmatizein*, shape, 679,10
skhesis, relation, 666,29.31; 672,14; *ta kata skhesin*, which exist relationally, 667,23; relational, 667,25
skopein, investigate, 653,16; examine, 657,17
smikrotês, smallness, 689,33-4
sôma, 645,29-1; 646,17.24; 647,8.19.22.33; 648,1-2.4.7.9.11-12.20.25; 649,6.18.23.24; 650,17-18.31; 651,3.5.13.29.30; 652,21.23.31; 653,2.25-30; 654,1-666,19 *passim*; 669,31; 670,9.11.12; 671,2; 672,27-8.32.39; 675,9.16.22.30; 676,1-3.6.32-3; 677,24.34; 678,4.18.35; 679,29.31-4.38; 680,13.32-3; 681,26; 682,3-7.15.22-4; 683,11.17.19.30.35; 684,7.15-16; 685,3.13.15.18.22.25-7.31.35.37; 686,2-4.6.8.11; 688,5.18.22; 689,7.12; 691,27-8.31; 692,1.4.9.11; 693,8.12-13.22-3.27.31.34-5; 694,1.2.9-10.12.17.23.27
sômatikos, bodily, 650,5; 652,8; 656,7-8.12.25; 657,9; 685,37; 690,21; 693,15.20; of body, 656,28
sphaira, ball, 671,5; 678,21
stadion, stadium, 672,5; *stadiaos*, of one stadium, 672,4
stasis, rest, 670,22
stereisthai, have not, 681,24-5; lack, 688,22
steremnios, thick, 678,12
stereos, solid, 693,24
sterêsis, privation, 647,4-5; 667,30; *en sterêsei*, be deprived, 664,9
stigmê, point, 672,23.27
stoikheion, letter, 672,31; element, 681,2
stoma, mouth, 659,24-5
strebloun, squeeze, 647,25.26; 655,3
strombos, top, 659,12
strongulos, rounded, 679,6
sullogismos, syllogism, 653,31; *hupothetikos*, hypothetical syllogism, 649,17
sullogizesthai, argue, 658,8

sumbainein, happen, 652,22-3; 659,27; 662,26; 665,27.33; 678,24; arise, 684,26; go on, 662,30; result, 676,35; 678,32; come about, 666,17; 675,14-15; 683,26
sumballein, contribute, 679,17
sumbebêkos, attribute, 681,28.34-5; 682,6.13-14
summetron, parallel, 657,16
summignunai, mingle, 669,2; mix with, 693,35; 694,21
sumpatheia, interaction, 694,7
sumperainein, conclude, 646,12-13; 686,33; 693,6
sumperasma, conclusion, 673,30
sumperilambanein, comprehend, 694,4
sumphônos, harmonise, 670,21
sumpilousthai, be compressed, 650,26; 660,10
sumpiptein, collapse, 650,27; 651,19-20
sumplêrousthai, determinate, 690,15
sumptôsis, collapse, 650,19
sunagein, reduce, 652,10; 689,21; bring about, 654,4; exhibit, 673,13-14; adduce, 656,1; collect, 666,7-8; draw conclusion, 671,15; 674,7.17; 682,2; conclude, 671,21; 673,21; 674,26; 675,13.22.25; 680,18; *sunagetai*, is conclusion, 674,32; follows, 675,12; 676,31
sunagôgê, contraction, 690,29
sunaition, cooperating cause, 664,6
sunanairein, also disprove, 667,2
sunanapherein, carry upwards with itself, 685,26
sunapôtheisthai, be carried away, 669,3-4
sunaptesthai, be together, 669,11
sundesmos, conjunction, 673,32
ou suneinai, depart, 653,36
sunekheia, continuity, 666,3; 691,32; 694,7; *kata tên sunekheian*, as continuous, 668,29
sunekhein, hold together, 671,9.11
sunekhes, continuous, 645,27.28; 648,12.19.21.25; 651,28; 659,12; 666,1.3.4; 668,30; 669,3; 684,15; 693,13; *sunekhôs*, continuously, 687,6.15
sunekhizein, be continuous, 657,24
sunekphainein, make clear also, 654,5
sunelkesthai, be pulled along, 652,24
sunêmmenon, consequent, 649,21-2; 650,1; consequence, 659,6; hypothetical, 686,20; *logos*, hypothetical argument, 660,14-15; 680,14; hypothetical premiss, 663,23
sunepôthein, add its pressure, 669,7
sunergein, do towards, 671,10-11
sunexistanai, be conjoined, 678,5
sungignoskein, recognise, 692,22
sunizanein, contract, 650,18; 651,15.21; congregate, 660,4; close up, 662,34
sunizêsis, congregation, 660,1
sunkhôrein, admit, 646,12; 657,3; allow, 688,3; concede, 684,31; agree, 693,10; permit, 647,30
sunkrinein, conjoin with, 686,9
sunodos, introduction, 694,15
sunoran, comprehend, 676,17
sunôthein, compress, 669,2; push together, 684,16; push, 694,21
sunteinein, lead up to, 676,15
sunthesis, composition, 650,23-4
sunthlibein, compress, 647,23; squeeze, 659,30
suntithenai, compose, 672,19-2.24.28
suntrekhein, be at the same time, 678,34
sustasis, consistency, 677,20; 678,15; composition, 694,14; figure, 689,20
sustellein, gather, 660,2-3; contract, 683,11.14.16; 689,35; decrease, 690,11
sustolê, decrease, 690,21

takhos, speed, 679,17; 684,11
taxis, be organised, 680,25
teinein, tend, 657,7; claim, 666,21
tekmêrion, confirmation, 650,20
tephra, ash, 648,28; 651,9.14-15.18.21; 662,14 *passim*; 663,1.2

terein, preserve, 684,1
thalassa, sea, 649,20; 683,25; 685,1
thaumastos, remarkable, 660,25
thatton, faster, 671,20.34; 672,3; 676,27; 678,25.29.33.39; 679,2.26; 680,1.3; 685,31; more quickly, 677,28.36; 678,26
theasthai, contemplate, 653,27
theôreisthai, be seen, 648,9; 692,34
theôria, science, 670,2
thermainesthai, grow hot, 658,31-2
thermos, hot, 688,8.15; 689,2-9.11; 690,5.14; warm, 681,23; heat, 662,26; *thermotês*, warmth, 690,6.32; 693,14.20
thesis, postulate, 645,32; being placed, 665,16; thesis, 646,21; 687,12; position, 665,17; 681,27
thura, door, 648,30-2.34; *para thuras apantan*, be beside the point, 652,30; miss the point, 661,7
titanos, lime, 662,27-8
tithenai, posit, 646,9.10.14; 653,3-4; 663,3.5; 664,4; 676,2; 680,21; 692,15; affirm, 652,32; lay down, 667,12; postulate, 646,17; assume, 670,32; set, 666,26; make, 648,18; place, 664,10.25; 665,15-16.21.27-8.36; 666,31; 671,14; 680,34; 681,2.15; 691,30; treat, 646,23; establish, 663,11; produce, 647,15; set out, 648,35; 649,4; 650,17; 652,18-19; 666,21; 682,12
to di' hou, medium, 671,29.37; 672,8; 674,7; 684,33
to ei esti, existence, 646,5; 653,16
to einai, existence, 694,6; to exist, 694,7
to en hoi, that in which, 664,4-5.26; 666,23-4.32
to huph' hou, by which (efficient cause), 684,34
tode ti, particular thing, 656,7; 657,9
topikôs, of place, 649,13; 664,33; 687,24; 691,15; spatial, 665,12; 684,32
topos, place, 645,23-647,31 *passim*; 649,8.18; 650,32.33; 652,20; 653,2.17.19.22.24.36; 655,10; 656,29-32; 657,7-8.10-11.14-16.31-2.35; 658,12.15.18.20.23; 659,3-666,19 *passim*; 667,20; 670,9.19; 678,3; 680,16-19; 681,22.26.30.32-3; 682,4-7.15.22-4; 683,9-10.20; 684,22; 685,7.13; 693,1.18; 694,8.28
treis, three, 658,33
trepein, turn, 654,6-7
trephein, feed, 660,16.28.29; 661,2.30-3; nourish, 651,6-7
trias, triad, 652,5
trophê, food, 651,3.6; 660,16.27.30.32; 661,7-662,11 *passim*
tropos, way, 646,12; 656,4; 684,24; 687,17; 692,2; 693,4.10

xulon, wood, 678,16

zêtein, enquire, 646,5.30; 661,3; 670,24; 679,24; 693,5; want to know, 679,5; investigate, 653,10-11; look for, 656,5; ask, 653,18; seek for, 657,1; 681,19; *zêtoumenon*, problem, 662,34
zôion, animal, 669,30.32.34; 670,2; 690,15-16

Subject Index

Important concepts not included in this index may be found in the synopsis at the end of the Introduction.

abstraction
  void conceived by abstraction, 692,21-3
accretion
  things grow not necessarily by accretion, 660,17-21
  and nourishment, 661,1-4.20-36
actual, actuality
  void in actuality, 692,13
  actual void, continuity and sympathy, 694,6-8
addition
  in nourishment, 661,13
  in growth, 690,30-4
  in rarefaction and condensation, 691,11-17
additional premiss, 663,23-7
Alexander of Aphrodisias (Peripatetic of the second century AD)
  on the jar receiving wine, 650,21
  on the Pythagorean doctrine of the void, 651,28-9
  on the 'what it signifies', 653,17-9
  on the full as object of touch, 654,34-655,2, 655,11-17
  his reading of 214a7, 655,22-7
  the air is only tangible, 655,29-32
  though three-dimensional, void cannot be body, 657,36-658,1
  on the argument from nourishment, 660,32-661,15
  interprets 214b21-2, 665,14-16
  on some copies of the *Physics*, 667,3-5
  on forced motion, 668,13-16
  criticises the Stoic theory of holding power, 671,4-15
  on 215b21-2, 673,28-31
  on the ratio of void to full, 675,2-11
  on the density of the media, 675,29-676,7.14-20
  his reading of 216a17, 678,30-1
  on the Atomist theory of falling atoms, 679,12-22.32-7
  on transformation, 684,3-4
  on mixture, 692,2-6
alteration
  void cannot cause alteration, 658,30-3; 659,5
  does not involve accretion, 660,18-25
  impossible if there is no rarefaction and condensation, 683,9-21
  caused by the matter, 692,29-34
Anaxagoras and his followers
  try to refute void by pointing out that air is something, 647,20-3; 652,28-32; 655,2-4.17-18; 656,20-1
Aspasius (Peripatetic of the second century AD)
  identifies the 'how it is' with the 'what it is', 646,34
Aristotle
  on Anaxagoras' argument against void, 647,23-32
  on the Pythagorean argument, 651,29-32
  on the arguments for void, 652,11-19
  reference to *An. Post.* 2.1, 89b24 ff., 653,9-10
  reference to *De Cael.* 1.3, 269.b29 ff., 655,36-7
  Alexander's reference to *Cat.* 6, 657,37-658,4

reference to *DA* 2.4, 415b26, 660,27-8
reference to *GC* 1.5, 320a8, 661,3-4
solves the argument from feeding, 661,24-6
on the argument from ash, 662,31-663,3
doubtful reference to *MA*, 669,35-670,1

attraction
Strato's argument, 652,21-2; 663,4-9
Plato seems to abolish it, 663,5-6

bodiless, incorporeal
those who eliminate bodiless nature, 653,26-7
and ash, 662,18-21
air is more of this sort than water, 673,2-4
light is such, 693,20-2

bodily
on being of this kind (Melissus), 650,4-6
Pythagorean view, 652,7-13
bodily substance as particular thing, 656,6-8.28; 657,8-11
bodily lightness, 685,37-686,1
bodily size, 690,20-1
bodily power, 693,13-22

boundless, infinite, unlimited
motion, 645,26-9
void as boundless, 663,15-17; 664,10-32; 667,18-23; 671,1-15
Aristotle's argument against such void, 682,1-3

cause
cooperating cause: void cannot be such, 664,6-7
efficient cause: void cannot be such, 664,3-6; 683,34
final cause: Plato's interpretation, 694,2-6

channel, pore, duct
in ash, 651,18-20
Strato's notion of attraction, 652,22-5
nourishment, 662,10-13
they are not empty, 692,9-11

coming to be
in moisture, 693,16
and alteration, 660,23-5; 687,12-13
Atomist theory expounded by Alexander, 679,20-2
dependence on rarefaction and condensation, 683,8-10

compression
argument based on it, 655,15-30
its refusal, 659,29-660,11
in rarefaction, 682,25-684,9
in motion, 686,18-687,32
Plato's notion (in *Tim.*), 694,13-16

concept, conception
of the void, 647,16-20; 648,1-7.31; 653,7-13; 656,6-7.28-35; 657,6; 694,29
of the separated void, 649,4-7
contrasted to hypostasis, 648,10
parallel to definition (*logos*), 688,12-13
common conception as criterion, 646,35-6

conceptual, akin to mathematical, 690,9-10

condensation
generally, 682,5-690,17
concerning the spread of light, 693,25-33

congregation, concerning empty pockets in bodies, 659,32-660,4

continuity
motion is continuous, 645,26-9
void destroys it, 648,12-15; 651,28-9; 691,32; 694,6-8
extracosmic void allows the cosmos continuous, 684,18-25
criticism of Melissus' theory, 659,4-15
on wholes and parts, 665,5-666,12
motion of projectiles, 668,29-669,5
on transformation, 687,5-7

cosmos, universe, the whole
held together by a cohesive force (Stoics), 671,6-12
bodily cosmos, 652,8-13
void outside and within the universe, 648,13-25; 683,33; 691,30
inhales the void, 651,25-8; 652,7-17

bulging, 683,23-5, 35; 686,19-20, 23; 687,26; 688,1
transformation, 684,5-9
Plato on its circle (*Tim.*), 694,3-6

Democritus and Leucippus
on the void between bodies, 648,11-13
on extracosmic void, 648,13-15.18-21
on atomic fall, 679,13-16
determinating accounts, 690,11-17
diminution
ash, 662,27-33
generally, 676,24-6
division
of motion into natural and forced, 667,11-13
of the medium by the travelling body, 678,26-7; 679,9-12.20-2; 680,1-8; 684,12-17
concerning excess, 672,16-20

enmattered
contrasted to mathematical and conceptual, 690,9-10
concerning light, 693,23
Epicurus
there is extracosmic void, 648,16-17
on falling atoms, 679,13-16.32-4
Eudemus (disciple of Aristotle)
on the argument from ash, quotation from his *Physics* 3, 662,24-31
exchange of place
eliminates the need for void, 659,7-21
only in bodies moving in circle, 683,29-31

food, feed, nourishment
argument from it for void, 650,30-651,9
refusal, 660,11-662,14
form
Forms, 650,10
of bodies, 656,24-9; 689,30-3; 690,8-16
growth (Alexander), 661,1-6
matter deprived of forms, 688,18-23
rare as form, 691,19-21
void and matter, 692,20-38

heavens, heavenly bodies
in void, 656,15
unmoved, 659,19-20
HERE
separation, 652,12-13
otherness as cause of the void HERE, 652,15-18

impulse, impetus
air, 655,1-2
nature causing local motion, 663,29-31; 678,14-28
in projectiles, 668,27-32
of moving bodies, 671,31; 678,35-679,3.29-32; 680,1-9
differences among bodies, 677,31-4
intelligible, contrasted to bodily and partial, 650,4-6
(be) interspersed, be scattered
void interspersed in bodies, 663,17-18; 682,24; 684,16-26; 691,31-2; 692,1-8
concerning compression, condensation and rarefaction, 683,16-21; 687,9-10

kinship
of shape to size and quantity in general, 689,16-19

limit, limited
of the container, 657,14-15; 665,3-7
of the interval, 667,20-2

mathematical
parallel to conceptual, 690,9-10
mathematical body, interval (Alexander), 658,1-6
mathematical surface, 665,4-5
matter
cause of motion?, 646,10-15; 691,14-15; 692,18-693,2.19
and void, 646,14-15; 656,26-657.3
mathematical interval, 658,1-2
growth, 661,1-5
substrate matter, 688,4-36; 689,2-690,34

matter susceptible to change (*pathêtikos*), 693,27
Melissus (Eleatic of the fifth century BC)
denies movement to what exists, 649,35-650,4; 658,33-659,2
his denial concerns the intelligible being, 650,4-11
Aristotle's criticism, 659,4-17
Metrodorus of Chios (Atomist, fourth century BC)
thinks void is outside the universe, 648,15
mix, mixture, admixture
void mixed in bodies, 663,17; 684,7-8; 686,8; 688,26-9
rarity and density, 688,26-9; 690,5-6; 692,2-5 (Alexander)

nature
moves as efficient cause, 684,34
Pythagorean view, 651,24-32
void distinguishes the nature of the numbers, 652,2-4
contribution to the local motion, 663,24-34; 678,32-3; 684,34; 685,7-20; 690,22-4
place as three-dimensional nature, 665,34; 691,19-20
of the matter, 689,15

otherness, diversity
belongs THERE, causes void HERE, 650,8-11; 652,15-17
Plato called not-being, 652,17-18
distinguishes the Forms THERE, 652,8

Parmenides
void has no room in what really exists, 650,11-13
portion of his Fr., B 7 D-K, 650,13
participate
concerning otherness, 652,9-10
vegetative soul, 660,29-30
limit, 665,5-6
particular thing, bodily substance, 656,6-8.28; 657,8-11
Peripatetic hypothesis concerning light, 693,19-22

Plato
on otherness as not-being, reference to *Sph*. 258B, 652,14-15
seems to abolish attraction, 663,6
on the resting earth, 666,23-6
quotation of *Phdo*. 109A, 666,25
on mutual replacement, 668,32-669,2
quotation from *Tim*. 58E-59A, 668,34-669,2
on space, *Tim*. 58A, 694,3-6; *Tim*. 58B, 694,13-16; *Tim*. 58E, 694,19-21
Platonic circles after Aristotle
void is always filled by bodies, 657,22-3
Porphyry
his version of 213a32-3, 648,17-18
on the early Atomists and Pythagoreans, 648,18-22
power
of the void in the numbers, 652,2-4
of attraction, 663,5-6
of the thrower, 669,5-9
holding power (Stoics), 671,8-11
concerning impetus, 677,3-4; 678,1
like the warmth, 693,14-20
projectiles (forced motion), 668,10-670,3
proportion, ratio
among motions through void and full, 673,5-678,5
concerning the time of travel through the void, 685,30-5; 686,1-11
push
forced motion, 668,13-669,16
motion as series of successive pushes, 687,15
Plato's view, 694,1-20
Pythagoreans
the universe inhales the void outside it, 648,16.21-2; 651,23.25-8
their argument for void, 649,6-7
Neopythagorean view of the void, 652,7-17

relation, relational
void, 666,26-31

of the full to the void, 672,12-14
relational beings change easily, 667,23-4
differences, 667,25
replacement, mutual
only in bodies moving in circle, 683,29-31; 687,18
in projectiles, 668,24-32
impossible in the void, 669,15-16
Plato's view, 694,16-21

sense, sensible
touch, 654,10-656,3
sensible volume, 689,33-690,11
Stoics (Chrysippus and Cleomedes)
on holding power, 671,8-11
Strato of Lampsacus (head of the Lyceum between 288/5-270/67 BC)
on the arguments for void, 652,19-25
on attraction, 663,3-8
proof for void from the percolation of light, 693,11-18
substance
bodies, 650,18
not only in volume, 660,7-8
of the ash, 662,31-2
bodily substance, 686,6-8, 28; 687,8-11
alteration, 692,32-7
subtraction
concerning curvature, 690,31-2
substrate
matter, 688,4-36; 689,2-690,34
the full and the place are the same in substrate, 646,26-7
the matter and the place are the same in substrate, 656,30-1
as medium, 674,16-19; 677,15-16; 678,18-26; 679,9-11; 680,8
sympathy
void destroys it, 694,6-8

tangible, touch
void as interval, 654,10-656,3; 657,8-9
Themistius (fourth century AD)
on transformation, 684,1-5
quotation of his *in Phys.*, 135,26-7, 684,2-3
THERE
otherness instead of void, 650,8-14; 652,8-17
time
ratios, 671,23-4; 672,3-678,8
timeless (instantaneous)
motion through void, 671,20-1; 672,11-12; 673,4-5.26-7; 677,11-34
transformation
in general, 686,17-687,33
concerning substrate, 687,35-688,24; 692,28-34

unseparated
Porphyry's reading and explanation, 684,18-26
void, 657,21-7; 658,22-4; 684,13-14; 686,13-14
intervals, 681,27-9

Xouthos (Pythagorean of the fifth century BC)
on the bulging universe, 683,23-6